Reliability Technology, Human Error, *and* Quality *in* Health Care

Reliability Technology, Human Error, *and* Quality *in* Health Care

B.S. Dhillon

CRC Press is an imprint of the
Taylor & Francis Group, an **informa** business

CRC Press
Taylor & Francis Group
6000 Broken Sound Parkway NW, Suite 300
Boca Raton, FL 33487-2742

First issued in paperback 2019

ISBN-13: 978-1-4200-6558-9 (hbk)
ISBN-13: 978-0-367-38775-4 (pbk)

Library of Congress Cataloging-in-Publication Data

Dhillon, B. S.
Reliability technology, human error, and quality in health care / author, B.S. Dhillon.
p. ; cm.
Includes bibliographical references and index.
ISBN 978-1-4200-6558-9 (hardcover : alk. paper)
1. Medical care--Quality control. 2. Reliability (Engineering) 3. Medical errors. 4. Medical instruments and apparatus--Safety measures. I. Title.
[DNLM: 1. Quality of Health Care. 2. Equipment Safety. 3. Medical Errors. 4. Models, Statistical. 5. Risk Management. W 84 AA1 D4996r 2007]

RA399.A1D487 2007
362.1--dc22 2007048968

Visit the Taylor & Francis Web site at
http://www.taylorandfrancis.com

and the CRC Press Web site at
http://www.crcpress.com

This book is affectionately dedicated to ancient Greek writers such as Herodotus, Diodorus, Plutarch, Strabo, Arrian, Ptolemy, and Ammianus Macellinus (a Roman writer of Greek ancestry), whose writings have helped me to trace my ancient Scythian ancestry, resulting in the publication of a book on this subject.

Contents

Preface

Each year a vast sum of money is spent on health care worldwide. For example, per the Congressional Budget Office's estimate, by 2008 total national health care expenditures in the United States will top $2 trillion. The current vast expenditures on health care still fail to deliver appropriate care because over 1,000 Americans die each week due to health-related problems, and more than $9 billion is lost in productivity annually.

The health care system is composed of many elements, including equipment and humans. In the effective functioning of such system reliability technology, human factors, and quality play an important role. Although, over the years, a large number of journal and conference proceedings articles on topics concerning health care have appeared, there are only a small number of books written on these topics. In fact, to the best of author's knowledge, no book covers all these topics within its framework. This causes a great deal of difficulty for information seekers, because they have to consult many different and diverse sources.

Thus, the main objective of this book is to combine all these three topics into a single volume, to eliminate the need to consult many different and diverse sources in obtaining desired information. The book contains a chapter on mathematical concepts and another chapter on introductory material on reliability technology, human error, and quality considered essential to understand materials presented in subsequent chapters.

The topics covered in the volume are treated in such a manner that the reader will require no previous knowledge to understand the contents. At appropriate places, the book contains examples along with their solutions and at the end of each chapter is a list of problems to test reader comprehension. The sources of most of the material presented are given in the reference section at the end of each chapter. A comprehensive list of references covering the period 1962 to 2006 on reliability technology, human error, and quality in health care is provided in the appendix, to give readers a view of the depth of developments in these areas.

This book is composed of 11 chapters. Chapter 1 presents the need for reliability technology, human error studies, and quality in health care; history; related facts, figures, terms, and definitions; and sources for obtaining useful information.

Chapter 2 reviews mathematical concepts considered useful to understand subsequent chapters. It covers topics such as arithmetic mean, standard deviation, Boolean algebra laws, probability properties, basic probability definitions, probability distributions, and Laplace transforms. Chapter 3 presents introductory aspects of reliability technology, human error, and quality.

Chapter 4 is devoted to medical device safety and quality assurance. Some of the topics covered in the chapter are medical device safety requirements, safety in device life cycle, safety analysis tools for medical devices, regulatory compliance of medical device quality assurance, and methods for medical device quality assurance.

Various important aspects of medical device software quality assurance and risk assessment are presented in Chapter 5.

Chapter 6 is devoted to medical device maintenance and sources for obtaining failure data. It covers topics such as medical equipment maintenance indexes, computerized maintenance management systems, mathematical models for maintenance, data sources, and organizations failure data. Chapter 7 presents various important aspects of human error in health care, including human error in medication, human error in medical devices, human error in anesthesia, and useful guidelines to prevent the occurrence of human error in health care.

Chapter 8 presents two important topics relating to human error in health care: health care human error reporting systems and models for predicting human reliability and error in health care. Chapter 9 is devoted to patient safety. Some of the topics covered in the chapter are goals, culture, measures and analysis methods, and organizations.

Chapter 10 covers various introductory aspects of quality in health care including reasons for escalating health care costs, health care quality goals, steps for improving quality, and quality indicators for use in hospital departments. Chapter 11 presents a number of quality methods considered useful for application in health care. Some of these methods are quality function deployment, control chart, Pareto chart, process flow diagram, force field analysis, scatter diagram, cause-and-effect diagram, and affinity diagram.

This book will be useful to many individuals, including health care professionals, administrators, and students; biomedical engineers and biomedical engineering graduate students, health care researchers and instructors, reliability, quality, human factors and safety professionals; and design engineers and associated professionals concerned with medical equipment.

The author is deeply indebted to many individuals, including colleagues, students, and friends, for their inputs and encouragement throughout the project. I thank my children Jasmine and Mark for their patience and intermittent disturbances that resulted in many desirable breaks. Last, but not least, I thank my boss, friend, and wife, Rosy, for typing various portions of this book and other related materials, and for her timely help in proofreading.

B.S. Dhillon
Ottawa, Ontario

About the Author

B.S. Dhillon is a professor of engineering management in the Department of Mechanical Engineering at the University of Ottawa, Ontario, Canada. He has served as chairman and director of the department/Engineering Management Program for more than 10 years at the same institution. He has published over 330 articles on reliability, safety, engineering management, and related subjects. He is or has been on the editorial boards of nine international scientific journals. In addition, Professor Dhillon has written 32 books on various aspects of reliability, design, safety, quality, and engineering management published by Wiley (1981), Van Nostrand (1982), Butterworth (1983), Marcel Dekker (1984), Pergamon (1986), etc. His books are used in over 75 countries and many are translated into German, Russian, and Chinese. He has served as general chairman of two international conferences on reliability and quality control held in Los Angeles and Paris in 1987.

Professor Dhillon has served as a consultant to various organizations and bodies and has many years of experience in the industrial sector. At the University of Ottawa, he has been teaching reliability, quality, engineering management, design, and related areas for over 27 years; and he has also lectured in over 50 countries, including keynote addresses at various international scientific conferences held in North America, Europe, Asia, and Africa. In March 2004, Professor Dhillon was a distinguished speaker at the Congressional conference and workshop on surgical errors (sponsored by the White House Health and Safety Committee and the Pentagon).

Professor Dhillon attended the University of Wales, where he earned a bachelor of science in electrical and electronic engineering and a master's of science in mechanical engineering. He received a PhD in industrial engineering from the University of Windsor.

1 Introduction

1.1 NEED FOR RELIABILITY TECHNOLOGY, HUMAN ERROR STUDIES, AND QUALITY IN HEALTH CARE

Each year a vast sum of money is being spent on health care worldwide and is increasing at a significant rate. For example, in 1980, the Organization for Economic Cooperation and Development (OECD) countries on the average spent around 4.2% of their gross domestic product (GDP) on health care and by 1984, the figure increased to approximately 8% of the GDP.[1] More specifically, since 1960, health care spending in the United States has increased from 5.3% of the gross national product (GNP) to almost 13% in 1991.[2]

Needless to say, according to the Congressional Budget Office's estimate, by 2008, total national health care expenditures in the United States will top $2 trillion. The current vast expenditures on health care still fail to deliver appropriate care because a large number of Americans die each week due to health-related problems, and more than $9 billion is lost in productivity annually.

The health care system is composed of many elements including equipment and humans. In the effective functioning of such a system, reliability technology, human factors, and quality play an important role. Therefore, there is a definite need for engineering and other professionals working in health care to have knowledge of reliability technology, human error, and quality in health care to a certain degree. In turn, this will result in better overall health care at lower cost.

1.2 HISTORY

An overview of historical developments in reliability technology, human error, and quality with respect to health care are presented below, separately.

1.2.1 Reliability Technology

Although the history of general reliability may be traced back to the early 1930s, the beginning of reliability in regard to health care can only be traced back to the latter part of the 1960s and is associated with medical equipment or devices. In the late 1960s, a number of publications appeared and covered topics such as "reliability of ECG instrumentation," "instrument induced errors in the electrocardiogram," "the effect of medical test instrument reliability on patient risks," and "safety and reliability in medical electronics."[3-7]

In 1980, a journal article presented a comprehensive list of publications related to medical equipment reliability.[8] In 1983, a book entitled *Reliability Engineering in Systems Design and Operation* devoted one entire chapter to medical equipment reliability.[9] In 2000, a book entitled *Medical Device Reliability and Associated Areas* covered various important aspects of medical equipment reliability and provided a comprehensive list of publications on the subject.[10]

Over the years, many other developments in reliability technology in health care have also taken place; and today, in the design of various medical systems/devices, reliability is considered seriously.

1.2.2 Human Error

Although the occurrence of human error in medicine may have been there ever since its first practice about 3,000 years ago, the earliest documented medical error-associated death in modern times may only be traced back to 1848.[11,12] It was associated with the administering of anesthetic.

It appears that the first serious efforts to study the occurrence of human error in health care were made in the late 1950s and the early 1960s.[13,14] They were also associated with anesthesia-related deaths. In 1994, a book entitled *Human Error in Medicine* appeared.[15] In this volume, many authors contributed chapters on different aspects of human error in medicine.

In 2003, a book entitled *Human Reliability and Error in Medical System* presented various aspects of human reliability and error in health care, including a number of methods used to study human error and reliability in general area, for application in medicine.[16] Also, the book contained a comprehensive list of publications on human error in medicine covering the period 1963 to 2000.

1.2.3 Quality

The history of quality in health care goes back to the 1860s, when Florence Nightingale (1820–1910), a British nurse, vigorously advocated the need for a uniform system to collect and evaluate hospital-related statistics.[17] Her efforts played an instrumental role in laying the foundation for health care quality assurance programs.

In 1914, in the United States E.A. Codman (1869–1940) studied the results of health care in regard to quality and emphasized issues such as the health and illness behaviors of the patients, the accreditation of institutions, the importance of licensure or certification of providers, and the economic barriers to receiving care, when examining the quality of care.[17,18]

In 2007, a book entitled *Applied Reliability and Quality: Fundamentals, Methods, and Procedures* presented a comprehensive list of references on quality in health care covering the period 1989 to 2005.[19] Over the years many other people have contributed to the topic, and nowadays, quality is considered very seriously by people and organizations in the health care system.

1.3 FACTS AND FIGURES

Some of the facts and figures, directly or indirectly, concerned with reliability technology, human error, and quality in health care are as follows:

- In 1997, the world market for medical devices was approximately $120 billion.[20]
- According to the World Health Organization (WHO), at least half of all medical equipment in many developing countries is unusable or partly usable, thus leading to an increased risk of harm to patients and health workers.[21,22]
- Approximately 10% of individuals who receive health care in industrialized countries are expected to suffer because of preventable harm and adverse events.[22-25]
- At any given point in time, over 1.4 million people around the world are suffering from infections acquired in hospitals.[25]
- Each day, approximately 150 people die in European Union (EU) countries due to hospital acquired infections.[26]
- A study reported that approximately 100,000 Americans die each year due to medical error.[27]
- The annual cost of medication errors is estimated to be over $7 billion to the United States economy.[28]
- In a study conducted by the Emergency Care Research Institute (ECRI), a sample of 15,000 hospital products were tested and it was found that 4% to 6% of the products were sufficiently dangerous to warrant immediate correction.[29]
- In the United States alone, the annual cost of hospital-based medication errors is around $2 billion.[30]
- In 1990, a study conducted by the U.S. Food and Drug Administration (FDA) revealed that approximately 44% of the quality-related problems that led to the voluntary recall of medical devices for the 6-year period October 1983 to September 1989 were the result of deficiencies or errors that could have been prevented through appropriate design controls.[31]
- A study reported that over 50% of technical-related medical equipment problems were caused by operators.[29]
- A study of critical incident reports, in an intensive care unit, from 1989 to 1999, reported that most of the incidents were the result of staff errors, not equipment failures.[32]
- A study of 37 million hospitalizations in the Medicare population in the United States, during the period 2000 to 2002, reported that there were approximately 1.14 millions patient safety incidents.[33]
- An Australian study revealed that during the period 1988 to 1996, 2.4% to 3.6% of all hospital admissions were drug-related, and approximately 32% to 69% were preventable.[34]
- A study of anesthetic incidents in operating rooms reported that 70% to 82% of the incidents were caused by humans.[35,36]
- In 1984, a study reviewed the records of 2.7 million patients discharged from New York hospitals and found that approximately 25% of the 90,609 patients who suffered an adverse event was the result of human error.[37]
- A Hong Kong teaching hospital administered a total of 16,000 anesthetics in one year and reported 125 related critical incidents.[38] A subsequent analysis of these incidents revealed that human error was an important factor in 80% of these cases.[38]

- A study of 5,612 surgical admissions to a hospital reported that 36 adverse outcomes were due to human error.[39]

1.4 TERMS AND DEFINITIONS

Many terms and definitions are used in the area of reliability technology, human error, and quality in health care. This section presents some of the more frequently used such terms and definitions taken from various sources.[27,40-49]

- **Adverse event.** This is an injury due to a medical intervention.
- **Anesthesiology.** This is a branch of medicine that deals with the processes of rendering patients insensible to various types of pain during surgery or when faced with acute/chronic pain states.
- **Failure.** This is the inability of an item to function within the specified guidelines.
- **Health care.** This is services provided to communities or individuals for monitoring, maintaining, restoring, or promoting health.
- **Health care organization.** This is an entity that provides, coordinates, and/or insures medical-associated services for people/public.
- **Human error.** This is the failure to perform a specified task (or the performance of a forbidden action) that could result in disruption of scheduled operations or damage to equipment and property.
- **Human factors.** This is a study of the interrelationships between humans, the tools they utilize or use, and the surrounding environment in which they live and work.
- **Human reliability.** This is the probability of carrying out a specified task successfully by humans at any required stage in system operation within the stated minimum time limit (if the time requirement is stated).
- **Maintainability.** This is the probability that a failed item will be restored/repaired to its satisfactory working condition.
- **Maintenance.** This is all measures necessary for retaining an item in, or repairing/restoring it to, a desired state; ensuring that physical assets continue to satisfy their specified missions.
- **Mean time to failure.** This is in the case of exponentially distributed times to failures, the sum of the operating time of given items over the total number of failures.
- **Medical device.** This is any machine, apparatus, instrument, implant, in vitro reagent, implement, contrivance, or other related or similar article, including any component, part, or accessory which is intended for application in diagnosing diseases or other conditions or in the treatment, mitigation, cure, or prevention of disease or intended to affect the structure or any function of the body.[49]
- **Medical technology.** This is equipment, methods, drugs, and procedures by professionals working in the area of health care, in delivering medical care to people and the systems within which such care is actually delivered.

- **Medication error.** This is any preventable event that may cause or result in incorrect medication use or patient harm while the medication is in the control of a patient, a health care professional, or a consumer.
- **Mission time.** This is the time during which the item is performing its specified mission.
- **Operator error.** This is an error that occurs when an equipment/system/item operator does not follow proper or correct procedures.
- **Patient safety.** This is freedom from accidental injury and ensuring patient safety involves the creation of operational processes/systems that lower the likelihood of error occurrence and enhance the likelihood of error occurrence interception.
- **Quality.** This is the extent to which the properties of a product or service produce/generate a stated outcome.
- **Quality assurance.** This is the measurement of the degree of care given or provided (assessment) and, when necessary, mechanisms for improving it.
- **Quality of care.** This is the level to which delivered health services meet established professional standards and judgments of value to all consumers.
- **Reliability.** This is the probability that an item will perform its specified function satisfactorily for the desired period when used according to the stated conditions.
- **Risk.** This is the degree of probability, chance, or possibility of loss.
- **Total quality management.** This is a philosophy of pursuing continuous improvement in all processes through the integrated efforts of all people associated with the organization.

1.5 USEFUL INFORMATION ON RELIABILITY TECHNOLOGY, HUMAN ERROR, AND QUALITY IN HEALTH CARE

There are many sources for obtaining, directly or indirectly, reliability technology, human error, and quality in health care-related information. Some of the most useful sources are presented below under different categories.

1.5.1 JOURNALS

American Journal of Medical Quality
British Journal of Anesthesia
British Medical Journal
Canadian Medical Journal
Drug Safety
IEEE Transactions on Reliability
Industrial Quality Control
International Journal of Healthcare Quality Assurance
International Journal of Reliability, Quality, and Safety Engineering
Journal of Professional Nursing
Journal of Quality Clinical Practice
Journal of Quality Technology
Journal of the American Medical Association
Journal of Risk and Reliability

The Lancet
Medical Device & Diagnostic Industry
Microelectronics and Reliability
New England Journal of Medicine
Quality and Reliability Management
Quality and Safety in Healthcare
Quality Progress
Quality Review
Reliability Engineering and System Safety
Reliability Review

1.5.2 Books

Blank, R. *The Basics of Reliability.* New York: Productivity Press, 2004.
Bogner, M.S., ed. *Human Error in Medicine.* Hillsdale, NJ: Lawrence Erlbaum Associates, 1994.
Dhillon, B.S. *Design Reliability: Fundamentals and Applications.* Boca Raton, FL: CRC Press, 1999.
Dhillon, B.S. *Human Reliability and Error in Medical System.* River Edge, NJ: World Scientific Publishing, 2003.
Dhillon, B.S. *Human Reliability with Human Factors.* New York: Pergamon Press, 1986.
Dhillon, B.S. *Medical Device Reliability and Associated Areas.* Boca Raton, FL: CRC Press, 2000.
Elsayed, E.A. *Reliability Engineering.* Reading, MA: Addison-Wesley, 1996.
Graham, N.O., ed. *Quality in Health Care.* Gaithersburg, MD: Aspen Publishers, 1995.
Grant Ireson, W., Coombs, C.F., Moss, R.Y., eds. *Handbook of Reliability Engineering and Management.* New York: McGraw-Hill, 1996.
Kales, P. *Reliability: For Technology, Engineering, and Management.* Upper Saddle River, NJ: Prentice Hall, 1998.
Kohn, L.T., Corrigan, J.M., Donaldson, M.S., eds. *To Err Is Human: Building a Safer Health System.* Washington, DC: National Academy Press, 1999.
McLaughlin, C.P., Kaluzny, A.D., eds. *Continuous Quality Improvement in Health Care.* Boston: Jones and Bartlett Publishers, 2006.
Mociver, R.M. *Medical Nightmares: The Human Face of Errors.* Toronto, Canada: Chestnut Publishing Group, 2002.
Rosenthal, M.M., Sutcliffe, K.M., eds. *Medical Error: What Do We Know? What Do We Do?* New York: John Wiley & Sons, 2002.
Smith, G.M. *Statistical Process Control and Quality Improvement.* Upper Saddle River, NJ: Prentice Hall, 2001.
Smith, R., Mobley, R.K. *Rules of Thumb for Maintenance and Reliability Engineers.* Burlington, MA: Elsevier/Butterworth-Heinemann, 2007.
Stamatis, D.H. *Total Quality Management in Healthcare.* Chicago: Irwin Professional Publishing, 1996.

1.5.3 Conference Proceedings

Proceedings of the Annual Human Factors Society Conferences
Proceedings of the Annual ISSAT International Conferences on Reliability and Quality in Design
Proceedings of the Annual Reliability and Maintainability Symposium
Proceedings of the European Organization for Quality Annual Conferences
Proceedings of the First Symposium on Human Factors in Medical Devices, 1999
Proceedings of the First Workshop on Human Error and Clinical Systems (HECS'99), 1999

Proceedings of the Second Annenberg Conference on Enhancing Patient Safety and Reducing Errors in Healthcare, 1998
Transactions of the American Society for Quality Control (Annual Conference Proceedings)

1.5.4 Organizations

American Hospital Association, 1 North Franklin, Chicago, Illinois 60606; 325 7th Street NW Washington, DC 20004
American Medical Association, 515 N. State Street, Chicago, Illinois 60610
American Society for Quality, 310 East Wisconsin Avenue, Milwaukee, Wisconsin 53201
Australian Medical Association, 42 Macquaire Street, Barton, ACT (Australian Capital Territory) 2600, Australia
British Medical Association, BMA House, Tavistock Square, London WC1H9JP, UK
Canadian Medical Association, 1867 Alta Vista Drive, Ottawa, Ontario K1G 3Y6, Canada.
Center for Devices and Radiological Health, 1390 Piccard Drive, Rockville, Maryland 20857
Emergency Care Research Institute, 5200 Butler Pike, Plymouth Meeting, Pennsylvania 19462
IEEE Reliability Society, 3 Park Avenue, 17th Floor, New York, New York 10016
Institute of Medicine, 500 Fifth Street NW, Washington, DC 20001
National Institute for Occupational Safety and Health, 395 E Street SW, Washington, DC 20201
Society for Maintenance and Reliability Professionals, 8201 Greensboro Drive, Suite 300, McLean, Virginia 22102
Society of Reliability Engineers, Buffalo Chapter, P.O. Box 631, Cheektowaga, New York 14225
World Health Organization, Avenue Appia 20, CH-1211, Geneva 27, Switzerland

1.6 PROBLEMS

1. Write an essay on the historical developments in reliability technology, human error, and quality with respect to health care.
2. List at least four most important facts and figures, directly or indirectly, concerned with reliability technology, human error, and quality in health care.
3. Define the following three terms:
   - Health care
   - Patient safety
   - Quality of care
4. What is the difference between adverse event and risk?
5. List six of the most important journals for obtaining reliability technology, human error, and quality in health care-related information.
6. List four of the most important books to obtain reliability technology, human error, and quality in health care-related information.
7. Define the following four terms:
   - Reliability
   - Human error
   - Quality
   - Quality assurance
8. What is medical technology?

9. List five of the most important organizations to obtain reliability technology, human error, and quality in health care-related information.
10. Write an essay on the need for reliability technology, human error studies, and quality in health care.

REFERENCES

1. Fuchs, M.C. Economics of U.S. Trade in Medical Technology and Export Promotion Activities of the U.S. Department of Commerce. In *The Medical Device Industry: Science, Technology, and Regulation in a Competitive Environment*, edited by N.F. Estrin, 917–928. New York: Marcel Dekker, 1990.
2. Gaucher, E.J., Coffey, R.J. *Total Quality in Healthcare: From Theory to Practice*. San Francisco: Jossey-Bass Publishers, 1993.
3. Crump, J.F. Safety and Reliability in Medical Electronics. Proceedings of the Annual Reliability and Maintainability Symposium, 1969, 320–330.
4. Taylor, E.F. The Effect of Medical Test Instrument Reliability on Patient Risks. Proceedings of the Annual Reliability and Maintainability Symposium, 1969, 328–330.
5. Meyer, J.L. Some Instrument Induced Errors in the Electrocardiogram. *Journal of the American Medical Association* 201 (1967): 351–358.
6. Johnson, J.P. Reliability of ECG Instrumentation in a Hospital. Proceedings of the Annual Reliability and Maintainability Symposium, 1967, 314–318.
7. Gechman, R. Tiny Flaws in Medical Design Can Kill. *Hospital Topics* 46 (1968): 23–24.
8. Dhillon, B.S. Bibliography of Literature on Medical Equipment Reliability. *Microelectronics and Reliability* 20 (1980): 737–742.
9. Dhillon, B.S. *Reliability Engineering in Systems Design and Operation*. New York: Van Nostrand Reinhold, 1983.
10. Dhillon, B.S. *Medical Device Reliability and Associated Areas*. Boca Raton, FL: CRC Press, 2000.
11. Beecher, H.K. The First Anesthesia Death and Some Remarks Suggested by It on the Fields of the Laboratory and the Clinic in the Appraisal of New Anesthetic Agents. *Anesthesiology* 2 (1941): 443–449.
12. Cooper, J.B., Newbower, R.S., Kitz, R.J. An Analysis of Major Errors and Equipment Failures in Anesthesia Management: Considerations for Prevention and Detection. *Anesthesiology* 60 (1984): 34–42.
13. Edwards, G., Morlon, H.J.V., Pask, E.A. Deaths Associated with Anesthesia: A Report on 1,000 Cases. *Anesthesia* 11 (1956): 194–220.
14. Cliffton, B.S., Hotten, W.I.T. Deaths Associated with Anesthesia. *British Journal of Anesthesia* 35 (1963): 250–259.
15. Bogner, M.S., ed. *Human Error in Medicine*. Hillsdale, NJ: Lawrence Erlbaum Associates, 1994.
16. Dhillon, B.S. *Human Reliability and Error in Medical System*. River Edge, NJ: World Scientific Publishing, 2003.
17. Graham, N.O. Quality Trends in Health Care. In *Quality in Health Care*, edited by N.O. Graham, 3–14. Gaithersburg, MD: Aspen Publishers, 1995.
18. Codman, E.A. The Product of the Hospital. *Surgical Gynecology and Obstetrics* 28 (1914): 491–496.
19. Dhillon, B.S. *Applied Reliability and Quality: Fundamentals, Methods, and Procedures*. London: Springer, 2007.
20. Murray, K. Canada's Medical Device Industry Faces Cost Pressures, Regulatory Reform. *Medical Device & Diagnostic Industry* 19 (1997): 30–39.

21. Issakov, A. Health Care Equipment: A WHO Perspective. In *Medical Devices: International Perspectives on Health and Safety*, edited by C.W.G. Von Grutting, 48–53. Amsterdam: Elsevier, 1994.
22. Donaldson, L., Philip, P. Patient Safety: A Global Priority. *Bulletin of the World Health Organization* (Ref. No. 04–015776) 82 (December 2004): 892–893.
23. Vincent, C., Neale, G., Woloshynowych, M. Adverse Events in British Hospitals: Preliminary Retrospective Record Review. *British Medical Journal* 322 (2001): 517–519.
24. Baker, G.R., Norton, P.G., Flintoft, V., et al. The Canadian Adverse Events Study: The Incidence of Adverse Events Among Hospital Patients in Canada. *Canadian Medical Association Journal* 179 (2004): 1678–1686.
25. *Global Patient Safety Challenge: 2005–2006*. Report, World Health Organization, Geneva, Switzerland, 2005.
26. Patient Safety. Health First Europe, Chaussee de Wavre 214D, 1050 Brussels, Belgium, 2007. Available online at www.healthfirsteurope.org/index.php?pid=82.
27. Kohn, L.T., Corrigan, J.M., Donaldson, M.S., eds. *To Err Is Human: Building a Safer Health System*. Institute of Medicine Report. Washington, DC: National Academy Press, 1999.
28. Wechsler, J. Manufacturers Challenged to Reduce Medication Errors. *Pharmaceutical Technology* February (2000): 14–22.
29. Dhillon, B.S. Reliability Technology in Healthcare Systems. Proceedings of the IASTED International Symposium on Computers and Advanced Technology in Medicine, Healthcare, and Bioengineering, 1990, 84–87.
30. Smith, D.L. Medication Errors and DTC Ads. *Pharmaceutical Executive* February (2000): 129–130.
31. Schwartz, A.P. A Call for Real Added Value. *Medical Industry Executive* February/March (1994): 5–9.
32. Wright, D. Critical Incident Reporting in an Intensive Care Unit. Report, Western General Hospital, Edinburgh, Scotland, UK, 1999.
33. Patient Safety in American Hospitals. Report, Health Grades, Inc., Golden, Colorado, July 2004.
34. Roughead, E.E., Gilbert, A.L., Primrose, J.G., Sansom, L.N. Drug-Related Hospital Admissions: A Review of Australian Studies Published 1988–1996. *Medical Journal of Australia* 168 (1998): 405–408.
35. Chopra, V., Bovill, J.G., Spierdijk, J., Koornneef, F. Reported Significant Observations During Anesthesia: Perspective Analysis Over an 18–month Period. *British Journal of Anesthesia* 68 (1992): 13–17.
36. Cook, R.I., Woods, D.D. Operating at the Sharp End: The Complexity of Human Error. In *Human Error in Medicine*, edited by M.S. Bogner, 225–309. Hillsdale, NJ: Lawrence Erlbaum Associates, 1994.
37. Leape, L.L. The Preventability of Medical Injury. In *Human Error in Medicine*, edited by M.S. Bogner, 13–25. Hillsdale, NJ: Lawrence Erlbaum Associates, 1994.
38. Short, T.G., O'Regan, A., Lew, J., Oh, T.E. Critical Incident Reporting in an Anesthetic Department Quality Assurance Programme. *Anesthesia* 47 (1992): 3–7.
39. Couch, N.P., et al. The High Cost of Low-Frequency Events: the Anatomy and Economics of Surgical Mishaps. *New England Journal of Medicine* 304 (1981): 634–637.
40. Fries, R.C. *Medical Device Quality Assurance and Regulatory Compliance*. New York: Marcel Dekker, 1998.
41. Omdahl, T.P., ed. *Reliability, Availability, and Maintainability (RAM) Dictionary*. Milwaukee, WI: ASQC Quality Press, 1988.
42. Glossary of Terms Commonly Used in Healthcare. Prepared by the Academy of Health, Suite 701–L, 1801 K St. NW, Washington, DC, 2004.

43. Graham, N.O., ed. *Quality in Healthcare: Theory, Application, and Evolution*. Gaithersburg, MD: Aspen Publishers, 1995.
44. Fries, R.C. *Reliable Design of Medical Devices*. New York: Marcel Dekker, 1997.
45. McKenna, T., Oliverson, R. *Glossary of Reliability and Maintenance Terms*. Houston, TX: Gulf Publishing Company, 1997.
46. Naresky, J.J. Reliability Definitions. *IEEE Transactions on Reliability* 19 (1970): 198–200.
47. Definitions of Effectiveness, Terms for Reliability, Maintainability, Human Factors, and Safety. MIL-STD-721. Washington, DC: U.S. Department of Defense.
48. Von Alven, W. H., ed. *Reliability Engineering*. Englewood Cliffs, NJ: Prentice Hall, 1964.
49. Federal Food, Drug, and Cosmetic Act, as Amended, Section 201(h). Washington, DC: U.S. Government Printing Office, 1993.

2 Basic Mathematical Concepts

2.1 INTRODUCTION

As in the developments of other fields of science and technology, mathematics has also played an important role in the development of reliability technology, quality, and human reliability and error fields. The history of mathematics may be traced back more than 2,000 years to the development of our current number symbols. The first evidence of the use of these symbols is found on stone columns erected by the Scythian emperor Asoka of India in 250 BC.[1]

The development of the field of probability is relatively new and its history may be traced back to a gambler's manual written by Girolamo Cardano (1501–1576). In this manual, he considered some interesting issues on probability.[1,2] However, the problem of dividing the winnings in a game of chance was solved by Blaise Pascal (1623–1662) and Pierre de Fermat (1601–1665), independently and correctly. In 1657, Christiaan Huygens (1629–1695) wrote the first formal treatise on probability based on the Pascal-Fermat correspondence.

Additional information on historical developments in mathematics including probability is available in refs. [1, 2]. This chapter presents basic mathematical concepts considered useful to understand subsequent chapters of this book.

2.2 RANGE, ARITHMETIC MEAN, AND STANDARD DEVIATION

There are certain characteristics of data that play an important role to describe the nature of a given set of data, thus making better decisions. This section presents three statistical measures considered useful to perform analysis of reliability, human error, and quality data in health care.[3-5]

2.2.1 RANGE

This is a quite useful measure of dispersion or variation. It may simply be described as the difference between the smallest and the largest values in a given data set.

Example 2.1

Assume that the maintenance department of a hospital is responsible for the satisfactory operation of various types of medical equipment. The number of monthly problems with this equipment over the period of the past 15 months were as follows:

10, 15, 6, 25, 19, 30, 18, 7, 20, 7, 8, 10, 5, 12, and 35.

Find the range for the above data values.

By examining the above given data values, we conclude that the smallest and the largest values are 5 and 35, respectively. Thus, the range of the given data values is:

$$\begin{aligned} R &= Highest\ data\ value - lowest\ data\ value \\ &= 35 - 5 \\ &= 30 \end{aligned}$$

2.2.2 Arithmetic Mean

This is expressed by

$$m = \frac{\sum_{j=1}^{n} DV_j}{n} \tag{2.1}$$

where

m is the mean value.
n is the number of data values.
DV_j is the data value j; for $j = 1, 2, 3, \ldots, n$.

Example 2.2

The inspection department of an organization manufacturing medical equipment inspected 10 identical medical systems and discovered 5, 9, 4, 6, 10, 7, 20, 16, 25, and 2 defects in each system. Calculate the average number of defects per medical system.

Using the given data values in Equation (2.1), we get:

$$\begin{aligned} m &= \frac{5+9+4+6+10+7+20+16+25+2}{10} \\ &= 10.4\ defects / medical\ system \end{aligned}$$

Thus, the average number of defects per medical system is 10.4. In other words, the arithmetic mean of the data values is 10.4 defects per medical system.

2.2.3 Standard Deviation

This is a widely used measure of dispersion of data in a given data set about the mean and is expressed by:

$$\sigma = \left[\frac{\sum_{j=1}^{n} \left(DV_j - \mu\right)^2}{n} \right]^{1/2} \tag{2.2}$$

where

μ is the mean value.
σ is the standard deviation.

Three properties of the standard deviation associated with the widely used normal probability distribution, presented subsequently in the chapter, are as follows:

- 99.73% of the all data values are included between $\mu - 3\,\sigma$ and $\mu + 3\,\sigma$.
- 95.45% of the all data values are included between $\mu - 2\,\sigma$ and $\mu + 2\,\sigma$.
- 68.27% of the all data values are included between $\mu - \sigma$ and $\mu + \sigma$.

Example 2.3

Calculate the standard deviation of the data set given in Example 2.2.

The mean value of the data set, calculated in Example 2.2, is:

$$\mu = m = 10.4 \text{ defects / system}$$

Using the above calculated value and the given data values in Equation (2.2) yields:

$$\sigma = \left[\frac{(5-10.4)^2+(9-10.4)^2+(4-10.4)^2+(6-10.4)^2+(10-10.4)^2}{10}\right.$$

$$\left.\frac{+(7-10.4)^2+(20-10.4)^2+(16-10.4)^2+(25-10.4)^2+(2-10.4)^2}{10}\right]^{1/2}$$

$$\sigma = \left[\frac{29.16+1.96+40.96+19.36+0.16+11.56+92.16+31.36+213.16+70.56}{10}\right]^{1/2}$$
$$= 22.59$$

Thus, the standard deviation of the data set given in Example 2.2 is 22.59.

2.3 BOOLEAN ALGEBRA LAWS AND PROBABILITY PROPERTIES

Boolean algebra is named after mathematician George Boole (1815–1864) and it plays an important role in probability theory and reliability-related studies. Some of its laws are as follows[6,7]:

Commutative Law

$$L \cdot M = M \cdot L \tag{2.3}$$

$$L + M = M + L \tag{2.4}$$

where

L is an arbitrary set or event.
M is an arbitrary set or event.
+ denotes the union of sets.

Dot (·) denotes the intersection of sets. Sometimes Equation (2.3) is written without the dot but it still conveys the same meaning.

Idempotent Law

$$M\,M = M \tag{2.5}$$

$$M + M = M \tag{2.6}$$

Associative Law

$$(L + M) + Z = L + (M + Z) \tag{2.7}$$

$$(L\,M)\,Z = L\,(M\,Z) \tag{2.8}$$

Distributive Law

$$L\,(M + Z) = L\,M + L\,Z \tag{2.9}$$

$$L + MZ = (L + M)\,(L + Z) \tag{2.10}$$

Absorption Law

$$L\,(L + M) = L \tag{2.11}$$

$$L + (L\,M) = L \tag{2.12}$$

Some of the important event-related probability properties are presented below.[8]

The probability of occurrence of an event, say M, is

$$0 \leq P(M) \leq 1 \tag{2.13}$$

The probability of the union of K independent events is given by:

$$P(M_1 + M_2 + M_3 + \ldots\ldots + M_K) = 1 - \prod_{i=1}^{K} \left(1 - P(M_i)\right) \tag{2.14}$$

where

$P(M_i)$ is the probability of occurrence of event M_i; i = 1, 2, 3, …, K.

The probability of an intersection of K independent events is given by

$$P(M_1\ M_2\ M_3 \ldots\ldots M_K) = P(M_1)P(M_2)P(M_3)\ldots\ldots P(M_K) \tag{2.15}$$

The probability of the union of K mutually exclusive events is:

$$P(M_1 + M_2 + M_3 + \ldots\ldots + M_K) = -\sum_{i=1}^{K} P(M_i) \tag{2.16}$$

The probability of occurrence and nonoccurrence of an event, say M, is always:

$$P(M) + P(\overline{M}) = 1 \tag{2.17}$$

where

$P(M)$ is the probability of occurrence of event M.

$P(\overline{M})$ is the probability of nonoccurrence of event M.

Example 2.4

Simplify Equation (2.14) for $K = 2$. Thus, for $K = 2$, Equation (2.14) yields:

$$\begin{aligned} P(M_1 + M_2) &= 1 - \prod_{i=1}^{2} \left(1 - P(M_i)\right) \\ &= P(M_1) + P(M_2) - P(M_1)P(M_2) \end{aligned} \tag{2.18}$$

Thus, Equation (2.18) is the simplified version of Equation (2.14) for $K = 2$.

2.4 BASIC PROBABILITY-RELATED DEFINITIONS

This section presents a number of probability-related definitions considered useful to perform reliability, quality, and human error studies in health care.

2.4.1 Probability Density Function

For a continuous random variable, the probability density function is defined by[8,9]:

$$f(t) = \frac{dF(t)}{dt} \tag{2.19}$$

where

t is time (i.e., a continuous random variable).
$F(t)$ is the cumulative distribution function.
$f(t)$ is the probability density function. In reliability work, it is known as the failure density function.

Example 2.5

Assume that the failure probability at time t (i.e., cumulative distribution function) of a part used in a medical equipment is expressed by:

$$F(t) = 1 - e^{-\lambda t} \tag{2.20}$$

where

$F(t)$ is the cumulative distribution function or the part failure probability at time t.
λ is the part failure rate.

Obtain an expression for the probability density function. More specifically, in this case the part failure density function.

By substituting Equation (2.20) into Equation (2.19), we get:

$$\begin{aligned} f(t) &= \frac{d\left(1 - e^{-\lambda t}\right)}{dt} \\ &= \lambda e^{-\lambda t} \end{aligned} \tag{2.21}$$

Thus, Equation (2.21) is the expression for the probability density function.

2.4.2 Cumulative Distribution Function

For a continuous random variable, the cumulative distribution function is expressed by[8]:

$$F(t) = \int_{-\infty}^{t} f(x)\,dx \tag{2.22}$$

where

x is a continuous random variable.
$f(x)$ is the probability density function.

For $t = \infty$ Equation (2.22) becomes:

$$\begin{aligned} F(\infty) &= \int_{-\infty}^{\infty} f(x)\,dx \\ &= 1 \end{aligned} \tag{2.23}$$

It means that the total area under the probability density curve is equal to unity.

Example 2.6

Prove, using Equation (2.21), the Equation (2.20) cumulative distribution function. Thus, by substituting Equation (2.21) into Equation (2.22) for $t \geq 0$, we get:

$$\begin{aligned} F(t) &= \int_{0}^{t} \lambda e^{-\lambda x} dx \\ &= 1 - e^{-\lambda t} \end{aligned} \tag{2.24}$$

Equations (2.20) and (2.24) are identical.

2.4.3 Expected Value

The expected value of a continuous random variable is defined by[8]:

$$E(t) = \int_{-\infty}^{\infty} t\, f(t)\,dt \tag{2.25}$$

where

$E(t)$ is the expected value or mean value of the continuous random variable t.

Example 2.7

For $t \geq 0$, find the mean value of the function expressed by Equation (2.21).

Thus, by substituting Equation (2.21) into Equation (2.25) we get:

$$E(t) = \int_0^\infty t \lambda e^{-\lambda t} dt = \frac{1}{\lambda} \tag{2.26}$$

Thus, for $t \geq 0$, the mean value of Equation (2.21) is given by Equation (2.26).

2.4.4 Variance

The variance, $\sigma^2(t)$, of a continuous random variable t is expressed by[8,9]:

$$\sigma^2(t) = E(t^2) - [E(t)]^2 \tag{2.27}$$

where

$$E(t^2) = \int_0^\infty t^2 f(t) dt, \quad for\ t \geq 0. \tag{2.28}$$

2.5 PROBABILITY DISTRIBUTIONS

Over the years, a large number of probability distributions have been developed.[10] This section presents a number of probability distributions considered useful for performing reliability technology, human error, and quality-related studies in health care.

2.5.1 Exponential Distribution

This is probably the most widely used probability distribution in reliability studies. Its probability density function is defined by:

$$f(t) = \lambda e^{-\lambda t}, \quad for \quad t \geq 0, \lambda > 0 \tag{2.29}$$

where

t is time.
$f(t)$ is the probability density function.
λ is the distribution parameter. In reliability work, it is known as the item failure rate.

By substituting Equation (2.29) into Equation (2.22), we get the following expression for cumulative distribution function:

$$\begin{aligned} F(t) &= \int_0^t \lambda e^{-\lambda t} dt \\ &= 1 - e^{-\lambda t} \end{aligned} \tag{2.30}$$

Using Equation (2.29) in Equation (2.25) yields the following expression for the distribution mean value:

$$\begin{aligned} m = E(t) &= \int_0^\infty t \lambda e^{-\lambda t} dt \\ &= \frac{1}{\lambda} \end{aligned} \tag{2.31}$$

where

m is the mean value.

2.5.2 Weibull Distribution

This distribution is named after W. Weibull, a Swedish mechanical engineering professor who developed it in the early 1950s.[11] The distribution probability density function is defined by:

$$f(t) = \frac{\beta t^{\beta-1}}{\theta^\beta} e^{-\left(\frac{t}{\theta}\right)^\beta}, \; t \geq 0, \; \beta \rangle 0, \; \theta \rangle 0 \tag{2.32}$$

where

β and θ are the distribution shape and scale parameters, respectively.

By substituting Equation (2.32) into Equation (2.22), we get the following cumulative distribution function:

$$\begin{aligned} F(t) &= \int_0^t \frac{\beta t^{\beta-1}}{\theta^\beta} e^{-\left(\frac{t}{\theta}\right)^\beta} dt \\ &= 1 - e^{-\left(\frac{t}{\theta}\right)^\beta} \end{aligned} \tag{2.33}$$

By using Equation (2.32) in Equation (2.25), we obtain the following equation for the distribution mean value:

$$m = E(t) = \int_0^t t \frac{\beta t^{\beta-1}}{\theta^{\beta}} e^{-\left(\frac{t}{\theta}\right)^{\beta}} dt$$
$$= \theta \Gamma\left(1 + \frac{1}{\beta}\right) \quad (2.34)$$

where

$\Gamma(\cdot)$ is the gamma function and is defined by

$$\Gamma(m) = \int_0^{\infty} t^{m-1} e^{-t} dt, \quad for \quad m > 0 \quad (2.35)$$

It is to be noted that for $\beta = 1$, the exponential distribution is the special case of this distribution. More specifically, for $\beta = 1$ and $\lambda = 1/\theta$, Equation (2.35) becomes Equation (2.29).

2.5.3 Normal Distribution

This is one of the most widely used probability distributions and sometimes it is called the Gaussian distribution, after a German mathematician named Carl Friedrich Gauss (1777–1855). However, the distribution was actually discovered by Abraham de Moivre (1667–1754) in 1733.[10]

The distribution probability density function is defined by:

$$f(t) = \frac{1}{\sigma\sqrt{2\Pi}} \exp\left[-\frac{(t-\mu)^2}{2\sigma^2}\right] dx, \text{ for } -\infty < t < +\infty \quad (2.36)$$

where

μ and σ are the distribution parameters (i.e., mean and standard deviation, respectively).

By substituting Equation (2.36) into Equation (2.22) we get:

$$F(t) = \frac{1}{\sigma\sqrt{2\Pi}} \int_{-\infty}^{t} \exp\left[-\frac{(x-\mu)^2}{2\sigma^2}\right] dx \quad (2.37)$$

Using Equation (2.36) in Equation (2.25) yields:

$$E(t) = \mu \quad (2.38)$$

2.5.4 General Distribution

This distribution is complex, but quite flexible. In reliability studies, it can be used to represent failure behavior of various different types of items including bathtub shape hazard rate (i.e., time-dependent failure rate) curve.[12] The probability density function of the distribution is defined by:

$$f(t)=\left[c\lambda\theta t^{\theta-1}+(1-c)\beta t^{\beta-1}\gamma e^{\gamma t^{\beta}}\right]\left[\exp\left[-c\lambda t^{\theta}-(1-c)\left(e^{\gamma t^{\beta}}-1\right)\right]\right] \tag{2.39}$$

$$\text{for } 0\leq c\leq 1 \text{ and } \lambda,\gamma,\theta,\beta>0.$$

where

t is time.
λ and γ are the distribution scale parameters.
β and θ are the distribution shape parameters.

By substituting Equation(2.39) into Equation (2.22), we get the following expression for the cumulative distribution function:

$$F(t)=1-\exp\left[-c\lambda t^{\theta}-(1-c)\left(e^{\gamma t^{\beta}}-1\right)\right] \tag{2.40}$$

The following are the special cases of this distribution:

- Bathtub shape hazard rate curve; for $\beta = 1$, $\theta = 0.5$
- Exponential distribution; for $c = 1$, $\theta = 1$
- Weibull distribution; for $c = 1$
- Rayleigh distribution; for $c = 1$, $\theta = 2$
- Makeham distribution; for $\beta = 1$, $\theta = 1$
- Extreme value distribution; for $c = 0$, $\beta = 1$

2.6 LAPLACE TRANSFORM DEFINITION, COMMON LAPLACE TRANSFORMS, FINAL-VALUE THEOREM, AND LAPLACE TRANSFORMS' APPLICATION IN SOLVING FIRST-ORDER DIFFERENTIAL EQUATIONS

The Laplace transform of a function, say $f(t)$, is defined by:

$$f(s)=\int_{0}^{\infty} f(t)e^{-st}dt \tag{2.41}$$

Table 2.1 Laplace Transforms of Some Selected Functions

$f(t)$	$f(s)$
e^{-at}	$\frac{1}{s+a}$
$te^{-}e^{-at}$	$\frac{1}{(s+a)^2}$
c (a constant)	$\frac{c}{s}$
$\frac{df(t)}{dt}$	$sf(s)-f(0)$
$\theta_1 f_1(t)+\theta_2 f_2(t)$	$\theta_1 f_1(s)+\theta_2 f_2(s)$
t	$\frac{1}{s^2}$
$\frac{f(t)}{t}$	$\int_s^{\infty} f(u)\,du$

where

t is a variable.
s is the Laplace transform variable.
$f(s)$ is the Laplace transform of $f(t)$.

Laplace transforms of selected functions that frequently occur in reliability studies are presented in Table 2.1.[13,14]

2.6.1 Laplace Transform: Final-Value Theorem

If the following limits exist, then:

$$\lim_{t\to\infty} f(t) = \lim_{s\to 0} s\,f(s) \tag{2.42}$$

Example 2.8

Prove by using the following equation that the left side of Equation (2.42) is equal to its right side:

$$f(t)=\frac{a_1}{(a_1+a_2)}-\frac{a_1}{(a_1+a_2)}e^{-(a_1+a_2)} \quad (2.43)$$

where

a_1 and a_2 are constants.

By substituting Equation (2.43) into the left side of Equation (2.42), we get:

$$\lim_{t\to\infty}\left[\frac{a_1}{(a_1+a_2)}-\frac{a_1}{(a_1+a_2)}e^{-(a_1+a_2)t}\right]=\frac{a_1}{a_1+a_2} \quad (2.44)$$

By using Table 2.1, we obtain the following Laplace transforms of Equation (2.43):

$$f(s)=\frac{a_1}{s(a_1+a_2)}-\frac{a_1}{(a_1+a_2)}\cdot\frac{1}{(s+a_1+a_2)} \quad (2.45)$$

Using Equation (2.45) in the right side of Equation (2.42) yields:

$$\lim_{s\to 0} s\left[\frac{a_1}{s(a_1+a_2)}-\frac{a_1}{(a_1+a_2)}\cdot\frac{1}{(s+a_1+a_2)}\right]=\frac{a_1}{(a_1+a_2)} \quad (2.46)$$

As the right sides of Equations (2.44) and (2.46) are the same, it proves that the left side of Equation (2.42) is equal to its right side.

2.6.2 Laplace Transforms' Application in Solving First-Order Differential Equations

In reliability studies, linear first-order differential equations are often solved. Laplace transforms are an effective tool to solve a set of linear first-order differential equations. The application of Laplace transforms to solve a set of first-order differential equations describing a reliability system is demonstrated through the following example.

Example 2.9

A repairable medical system can either be in two states: operating normally or failed. The following two differential equations describe the system:

$$\frac{dP_0(t)}{dt}+\lambda P_0(t)=P_1(t)\theta \quad (2.47)$$

$$\frac{dP_1(t)}{dt}+\theta P_1(t)=P_0(t)\lambda \tag{2.48}$$

At $t = 0$, $P_0(0) = 1$ and $P_1(0) = 1$.

where

$P_j(t)$ is the probability that the medical system is in state j at time t; for $j = 0$ (operating normally), $j = 1$ (failed).
λ is the medical system failure rate.
θ is the medical system repair rate.

Find solutions to Equations (2.47) and (2.48) by using Laplace transforms. By taking the Laplace transforms of Equations (2.47) and (2.48) and then using the initial conditions, we obtain:

$$P_0(s)=\frac{s+\theta}{s(s+\theta+\lambda)} \tag{2.49}$$

$$P_1(s)=\frac{\lambda}{s(s+\theta+\lambda)} \tag{2.50}$$

Taking the inverse Laplace transforms of Equations (2.49) and (2.50), we get:

$$P_0(t)=\frac{\theta}{(\lambda+\theta)}+\frac{\lambda}{(\lambda+\theta)}e^{-(\lambda+\theta)t} \tag{2.51}$$

$$P_1(t)=\frac{\lambda}{(\lambda+\theta)}-\frac{\lambda}{(\lambda+\theta)}e^{-(\lambda+\theta)t} \tag{2.52}$$

Equations (2.51) and (2.52) are the solutions to Equations (2.47) and (2.48).

2.7 PROBLEMS

1. Write an essay on the historical development in mathematics.
2. Prove that the left-hand side of Equation (2.10) is equal to its right hand side.
3. Simplify Equation (2.14) for $K = 3$.
4. What is the difference between independent events and mutually exclusive events?

5. What are the special case probability distributions of the general distribution?
6. Obtain, using the right-hand side of Equation (2.42), steady state values for Equations (2.49) and (2.50).
7. Prove that the sum of Equations (2.49) and (2.50) is equal to $1/s$. Comment on the end result.
8. Obtain Laplace transform, using Equation (2.41), of the following function:

$$f(t) = t\,e^{-at} \tag{2.53}$$

where
a is a constant.
t is a variable.

9. Prove Equation (2.38) by using Equation (2.36).
10. Mathematically define the following two items:
    - Cumulative distribution function
    - Variance
11. Prove Equation (2.40) by using Equation (2.39).

REFERENCES

1. Eves, H. *An Introduction to the History of Mathematics.* New York: Holt, Rinehart, and Winston, 1976.
2. Owen, D.B., ed. *On the History of Statistics and Probability.* New York: Marcel Dekker, 1976.
3. Spiegel, M.R, *Probability and Statistics.* New York: McGraw-Hill, 1975.
4. Spiegel, M.R. *Statistics.* New York: McGraw-Hill, 1961.
5. Dhillon, B.S. *Reliability, Quality, and Safety for Engineers.* Boca Raton, FL: CRC Press, 2004.
6. Fault Tree Handbook. Report No. NUREG-0492. Washington, DC: U.S. Nuclear Regulatory Commission, 1981.
7. Lipschutz, S. *Set Theory.* New York: McGraw-Hill, 1964.
8. Mann, N.R., Schafer, R.E., Singpurwalla, N.D. *Methods for Statistical Analysis of Reliability and Life Data.* New York: John Wiley & Sons, 1974.
9. Shooman, M.L. *Probabilistic Reliability: An Engineering Approach.* New York: McGraw-Hill, 1968.
10. Patel, J.K., Kapadia, C.H., Owen, D.B. *Handbook of Statistical Distributions.* New York: Marcel Dekker, 1976.
11. Weibull, W. A Statistical Distribution Function of Wide Applicability. *Journal of Applied Mechanics* 18 (1951): 293-297.
12. Dhillon, B.S. A Hazard Rate Model. *IEEE Transactions on Reliability* 29 (1979): 150.
13. Oberhettinger, F., Badii, L. *Tables of Laplace Transforms.* New York: Springer-Verlag, 1973.
14. Spiegel, M.R. *Laplace Transforms.* New York: McGraw-Hill, 1965.

3 Introduction to Reliability Technology, Human Error, and Quality

3.1 INTRODUCTION

Reliability technology is playing an important role during the design of engineering systems/products as our daily lives and schedules are becoming increasingly dependent on the satisfactory functioning of these systems/products. Some examples of these systems/products are automobiles, computers, trains, aircraft, space satellites, and buses. Normally, the required reliability of such systems/products is stated in their design specifications and then every effort is made during the design phase to meet this requirement satisfactorily.

Humans play an important role during the system life cycle: design and development phase, production phase, and deployment phase. Although the degree of the role may vary from one system phase to another, this role is subject to deterioration as a result of human error. According to one study, human error is the cause for 20% to 50% of all equipment failures.[1]

The importance of quality in business and industry is increasing rapidly throughout the world. Today, some of the quality-related challenges facing the industry are an alarming rate of increase in customer's quality requirements, Internet economy, need for improvements in methods and practices associated with quality-related activities, and high cost of quality. The cost of quality control activities accounts for approximately 7% to 10% of the sales revenues of manufacturers.[2]

This chapter presents various introductory aspects of reliability technology, human error, and quality considered useful to understand subsequent chapters of this book.

3.2 BATHTUB HAZARD RATE CURVE

The bathtub hazard rate curve is a concept widely used to represent failure behavior of engineering systems/products because their failure rates change with time. It is called bathtub hazard rate curve because its shape resembles the shape of a bathtub (Figure 3.1). The curve is divided into three regions: region I: burn-in period, region II: useful life period, and region III: wear-out period. During the burn-in period, the product/system hazard rate decreases with time. Some of the reasons for the occurrence of failures during this period are presented in Table 3.1.[3] Other names used for the burn-in period are "break-in period," "infant mortality period," and "debugging period."

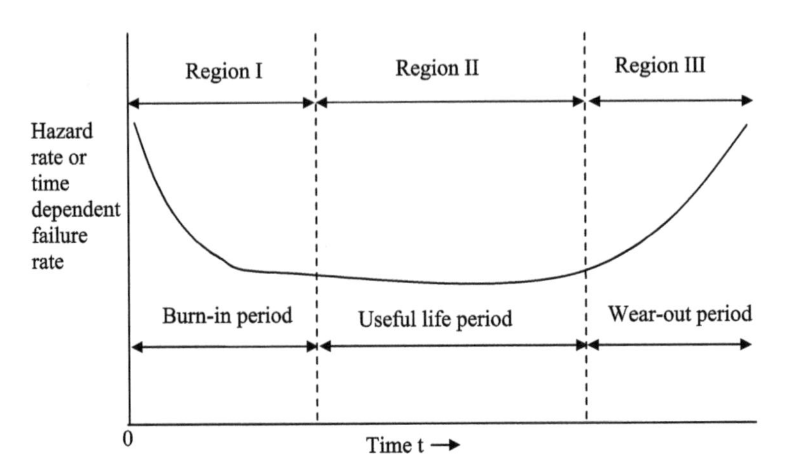

FIGURE 3.1 Bathtub hazard rate curve.

Table 3.1 Reasons for the Occurrence of Failures During Bathtub Hazard Rate Curve Regions I, II, and III

Bathtub Hazard Rate Curve Region	Reasons for Occurrence of Failures
I. Burn-in period	Poor quality, Human error, Poor workmanship, Substandard parts and materials, Poor debugging, Poor processes, Poor manufacturing methods, Incorrect packaging, Poor handling methods, and Wrong installation and start-up
II. Useful life period	Higher random stress than expected, Low safety factors, Human errors, Undetectable defects, Abuse, and Unavoidable conditions
III. Wear-out period	Wear due to aging, Wear due to friction, Incorrect overhaul practices, Short designed-in life of the product, Poor maintenance, and Corrosion and creep

During the useful life period, the system/product hazard rate remains constant and the failures occur unpredictably or randomly. Some of the reasons for their occurrence are also presented in Table 3.1.[3]

Finally, during the wear-out period, the system/product hazard rate increases with time. Some of the reasons for the occurrence of failures during this period are also presented in Table 3.1.[3]

3.3 COMMONLY USED FORMULAS IN RELIABILITY ANALYSIS

Three parameters widely used in reliability analysis are reliability, hazard rate (i.e., failure rate), and mean time to failure. General formulas to obtain these three parameters are presented below, separately.

3.3.1 General Formula for Reliability

This is expressed by[4]:

$$R(t) = e^{-\int_0^t \lambda(t)\,dt} \tag{3.1}$$

where

$R(t)$ is the reliability at time t.
$\lambda(t)$ is the hazard rate or time-dependent failure rate.

Example 3.1

Assume that the hazard rate of an x-ray machine is expressed by:

$$\lambda(t) = \lambda \tag{3.2}$$

where

λ is the x-ray machine failure rate.

Obtain an expression for the x-ray machine reliability.

By substituting Equation (3.2) into Equation (3.1), we get:

$$\begin{aligned} R(t) &= e^{-\int_0^t \lambda\,dt} \\ R(t) &= e^{-\lambda t} \end{aligned} \tag{3.3}$$

Thus, Equation (3.3) is the expression for the x-ray machine reliability.

3.3.2 General Formula for Hazard Rate

The general formula for hazard rate can be expressed in three different ways, as follows[4]:

$$\lambda(t) = \frac{f(t)}{R(t)} \tag{3.4}$$

or

$$\lambda(t) = \frac{f(t)}{1 - \int_0^t f(t)\,dt} \tag{3.5}$$

or

$$\lambda(t) = -\frac{1}{R(t)} \frac{dR(t)}{dt} \tag{3.6}$$

where

$f(t)$ is the failure density function.

Example 3.2

Prove by using Equations (3.3) and (3.6) that the x-ray machine's hazard rate is expressed by Equation (3.2).

Thus, by substituting Equation (3.3) into Equation (3.6), we get:

$$\begin{aligned} \lambda(t) &= -\frac{1}{e^{-\lambda t}} \cdot \frac{de^{-\lambda t}}{dt} \\ &= \lambda \end{aligned} \tag{3.7}$$

Equations (3.7) and (3.2) are identical.

3.3.3 General Formula for Mean Time to Failure

The general formula for mean time to failure can be expressed in three different ways, as follows[4]:

$$MTTF = \int_0^\infty R(t)\, dt \tag{3.8}$$

or

$$MTTF = \int_0^\infty t\, f(t)\, dt \tag{3.9}$$

or

$$MTTF = \lim_{s \to 0} R(s) \tag{3.10}$$

where

$MTTF$ is the mean time to failure.
s is the Laplace transform variable.
$R(s)$ is the Laplace transform of the $R(t)$.

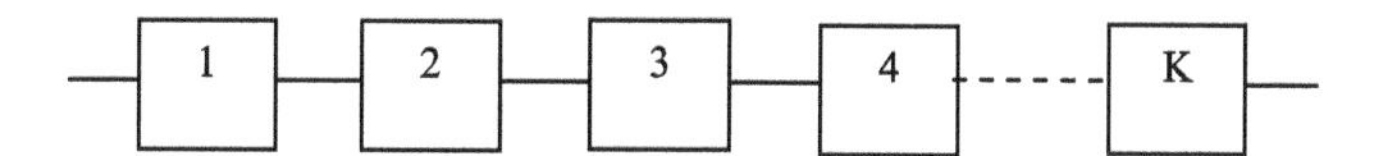

FIGURE 3.2 Block diagram of a series system containing K units.

Example 3.3

Obtain by using Equations (3.3) and (3.8) mean time to failure of the x-ray machine.

By substituting Equation (3.3) into Equation (3.8) we obtain:

$$MTTF = \int_0^\infty e^{-\lambda t}\,dt = \frac{1}{\lambda} \tag{3.11}$$

The mean time to failure of the x-ray machine is expressed by Equation (3.11).

3.4 RELIABILITY CONFIGURATIONS

Engineering systems can form various different configurations in performing reliability analysis. Some of the commonly occurring such configurations are presented below, separately.

3.4.1 Series Configuration

This is probably the most commonly occurring reliability configuration in engineering systems. Its block diagram is shown in Figure 3.2. Each block in the diagram denotes a unit/component. More specifically, the Figure 3.2 diagram represents a system composed of K units in series. In this case, if any one of the units fails, the system fails. In other words, all units of the system must operate normally for its successful operation.

If we let E_j denote the event that the jth unit is successful, then the Figure 3.2 series system reliability is expressed by[5]:

$$R_{SS} = P(E_1\, E_2\, E_3 \ldots . E_K) \tag{3.12}$$

where

$P(E_1\, E_2\, E_3 \ldots . E_K)$ is the probability of occurrence of events $E_1\, E_2\, E_3 \ldots . E_K$.
R_{SS} is the series system reliability.

For independent units, Equation (3.12) becomes:

$$R_{SS} = P(E_1)\; P(E_2)\, P(E_3) \ldots \ldots . P(E_K) \tag{3.13}$$

where

$P(E_j)$ is the occurrence probability of event E_j; for $j = 1, 2, 3, ..., K$.

If we let $R_j = P(E_j)$ for $j = 1, 2, 3, ..., K$ in Equation (3.13) becomes:

$$R_{SS} = R_1 R_2 R_3 ....R_K \tag{3.14}$$

where

R_j is the unit j reliability; for $j = 1, 2, 3, ..., K$.

For constant failure rate, λ_j, of unit j, using Equation (3.3), the reliability of the unit j is given by:

$$R_j(t) = e^{-\lambda_j t} \tag{3.15}$$

where

$R_j(t)$ is the reliability of unit j at time t.

Thus, by substituting Equation (3.15) into Equation (3.14), we get:

$$R_{SS}(t) = e^{-\sum_{j=1}^{K} \lambda_j t} \tag{3.16}$$

where

$R_{SS}(t)$ is the series system reliability at time t.

Using Equation (3.16) in Equation (3.8) yields the following expression for the series system mean time to failure:

$$\begin{aligned} MTTF_{SS} &= \int_0^{\infty} e^{-\sum_{j=1}^{K} \lambda_j t} \, dt \\ &= \frac{1}{\sum_{j=1}^{K} \lambda_J} \end{aligned} \tag{3.17}$$

where

$MTTF_{SS}$ is the series system mean time to failure.

By substituting Equation (3.16) into Equation (3.6) we get the following expression for the series system hazard rate:

$$\lambda_{SS} = \frac{1}{e^{-\sum_{j=1}^{K} \lambda_j t}} \left(-\sum_{j=1}^{K} \lambda_j \right) e^{-\sum_{j=1}^{K} \lambda_j t} = \sum_{j=1}^{K} \lambda_J \tag{3.18}$$

Example 3.4

Assume that a piece of medical equipment is composed of four independent subsystems. The failure rates of subsystems 1, 2, 3, and 4 are 0.0001 failures/hour, 0.0002 failures/hour, 0.0003 failures/hour, and 0.0004 failures/hour, respectively. Calculate the medical equipment reliability for a 50-hour mission and mean time to failure.

Using the above specified data values in Equation (3.16) yields:

$$R_{SS}(50) = e^{-(0.0001+0.0002+0.0003+0.0004)(50)} = 0.9512$$

By inserting the given data into Equation (3.17) we get:

$$MTTF_{SS} = \frac{1}{(0.0001+0.0002+0.0003+0.0004)} = 1{,}000 \ hours$$

It means that the medical equipment reliability and mean time to failure are 0.9512 and 1,000 hours, respectively.

3.4.2 Parallel Configuration

In this case, all units operate simultaneously and at least one of these units must operate normally for successful system operation. The parallel system or configuration, containing K units, block diagram is shown in Figure 3.3. Each block in the diagram represents a unit.

If we let $\overline{E_j}$ denote the event that the jth unit is unsuccessful, then the Figure 3.3 parallel system failure probability is expressed by[5]:

$$F_{ps} = P\left(\overline{E_1}\,\overline{E_2}\,\overline{E_3} \ldots \overline{E_K}\right) \tag{3.19}$$

where

$$P\left(\overline{E_1}\,\overline{E_2}\,\overline{E_3}\ldots.\overline{E_K}\right)$$

is the occurrence probability of failure events

$$\overline{E_1}\,\overline{E_2}\,\overline{E_3}\ldots.\text{ and }\overline{E_K}$$

F_{ps} is the parallel system probability of failure.

For independently failing units, Equation (3.19) becomes:

$$F_{ps}=P\left(\overline{E_1}\right)P\left(\overline{E_2}\right)P\left(\overline{E_3}\right)\ldots\ldots P\left(\overline{E_K}\right) \tag{3.20}$$

where

$P\left(\overline{E_j}\right)$ is the probability of occurrence of failure event $\overline{E_j}$; for j = 1, 2, 3, …., K.

If we let $F_j = \mathrm{P}\left(\overline{E_j}\right)$ for j = 1, 2, 3, …., K in Equation (3.20) becomes:

$$F_{ps}=F_1\,F_2\,F_3\ldots.F_K \tag{3.21}$$

where

F_j is the probability of failure of unit j; for j = 1, 2, 3, …., K.

By subtracting Equation (3.21) from unity, we get the following expression for the parallel system reliability:

$$R_{ps}=1-F_{ps}=F_1\,F_2\,F_3\ldots.F_K \tag{3.22}$$

where

R_{ps} is the parallel system reliability.

For constant failure rate, λ_j, of unit j, subtracting Equation (3.15) from unity and then inserting it into Equation (3.22) yields:

$$R_{ps}(t)=1-\left(1-e^{-\lambda_1 t}\right)\left(1-e^{-\lambda_2 t}\right)\left(1-e^{-\lambda_3 t}\right)\ldots\ldots\left(1-e^{-\lambda_K t}\right) \tag{3.23}$$

where

R_{ps} (t) is the parallel system reliability at time t.

For identical units, substituting Equation (3.23) into Equation (3.8) yields:

$$MTTF_{PS} = \int_0^{\infty} \left[1 - \left(1 - e^{-\lambda t}\right)^K\right] dt = \frac{1}{\lambda} \sum_{j=1}^{K} \frac{1}{j} \tag{3.24}$$

where

λ is the unit constant failure rate.
$MTTF_{ps}$ is the parallel system mean time to failure.

Example 3.5

Assume that a medical system is composed of two active, independent, and identical units and at least one of the units must operate normally for system success. The unit failure rate is 0.0008 failures/hour. Calculate the medical system reliability for a 60-hour mission and mean time to failure.

By substituting the given data into Equation (3.23), we get:

$$R_{ps}(60) = 1 - \left[1 - e^{-(0.0008)(60)}\right]\left[1 - e^{-(0.0008)(60)}\right] = 0.9978$$

Inserting the specified data values into Equation (3.24) yields:

$$MTTF_{ps} = \frac{1}{(0.0008)} \sum_{j=1}^{2} \frac{1}{j} = 1{,}875 \text{ } hours$$

Thus, the medical system reliability and mean time to failure are 0.9978 and 1,875 hours, respectively.

3.4.3 *m*-out-of-*K* Configuration

In this case all K units operate simultaneously and at least m units out of these K units must operate normally for system success. The series and parallel configurations of Sections 3.4.1 and 3.4.2 are the special cases of this configuration for $m = K$ and $m = 1$, respectively.

For independent and identical units, using the binomial distribution, the m-out-of-K configuration reliability is given by[5]:

$$R_{m/k} = \sum_{j=m}^{K} \binom{K}{j} R^j (1 - R)^{K-j} \tag{3.25}$$

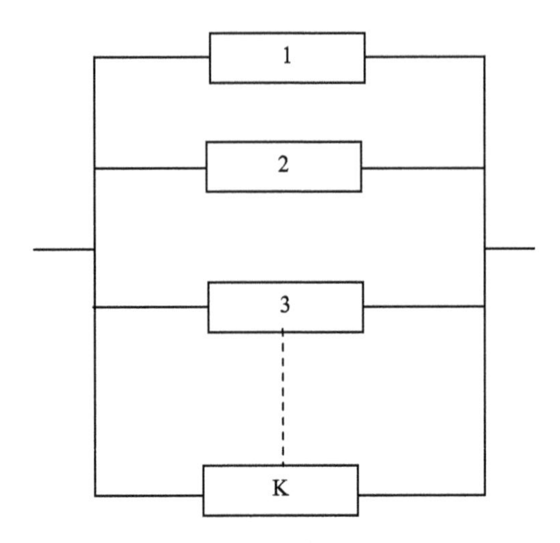

FIGURE 3.3 Block diagram of a parallel system containing K units.

where

$$\binom{K}{j} = \frac{K!}{(K-j)!\,j!} \tag{3.26}$$

R is the unit reliability.
$R_{m/k}$ is the m-out-of-K configuration reliability.

For constant failure rate, λ, of each unit, using Equations (3.3) and (3.25), we get:

$$R_{m/k}(t) = \sum_{j=m}^{K} \binom{K}{j} e^{-j\lambda t}\left(1-e^{-\lambda t}\right)^{K-j} \tag{3.27}$$

where

$R_{m/k}(t)$ is m-out-of-K configuration reliability at time t.

By substituting Equation (3.27) into Equation (3.8) we get:

$$\begin{aligned} MTTF_{m/k} &= \int_0^{\infty}\left[\sum_{j=m}^{K} \binom{K}{j} e^{-j\lambda t}\left(1-e^{-\lambda t}\right)^{K-j}\right]dt \\ &= \frac{1}{\lambda}\sum_{j=m}^{K}\frac{1}{j} \end{aligned} \tag{3.28}$$

where

$MTTF_{m/k}$ is the mean time to failure of the m-out-of-K configuration or system.

Example 3.6

A medical system is made up of four independent, identical, and active subsystems. At least three subsystems must operate normally for system success. The failure rate of a subsystem is 0.0002 failures per hour. Calculate the medical system mean time to failure.

By substituting the given data values into Equation (3.28) we get:

$$MTTF_{3/4} = \frac{1}{(0.0002)} \sum_{j=3}^{4} \frac{1}{j}$$
$$= \frac{1}{(0.0002)} \left(\frac{1}{3} + \frac{1}{4} \right)$$
$$= 2{,}916.7 \ hours$$

Thus, the medical system mean time to failure is 2,916.7 hours.

3.4.4 Standby System

This is another type of configuration used to improve system reliability. In this case, the system is composed of $(K + 1)$ units and only one unit operates and the remaining K units are kept in their standby mode. As soon as the operating unit fails, the switching mechanism detects the failure and then replaces the failed unit with one of the standby units. The system fails when all standby units fail.

For independent and identical units, time dependent unit failure rate, and perfect switching mechanism and standby units, the standby system reliability is expressed by[5]:

$$R_{Sd}(t) = \sum_{j=0}^{K} \left[\left[\int_0^t \lambda(t)\,dt \right]^j e^{-\int_0^t \lambda(t)\,dt} \right] / j! \tag{3.29}$$

where

K is the number of standby units.
$R_{Sd}(t)$ is the standby system/configuration reliability at time t.

For constant unit failure rate (i.e., $\lambda(t) = \lambda$), Equation (3.29) becomes:

$$R_{Sd}(t) = \sum_{j=0}^{K} \left[(\lambda t)^j e^{-\lambda t} \right] / j! \tag{3.30}$$

Using Equation (3.30) in Equation (3.8) yields:

$$MTTF_{Sd} = \int_0^{\infty} \left[\left[\sum_{j=0}^{K} (\lambda t)^j e^{-\lambda t} \right] / j! \right] dt$$
$$= \frac{K+1}{\lambda} \tag{3.31}$$

Example 3.7

A standby medical system has two independent and identical units: one operating and the other on standby. The switching mechanism to replace the failed unit is perfect and the standby unit remains as good as new in its standby mode. The constant failure rate of each unit is 0.0005 failures per hour. Calculate the standby medical system reliability for a 100-hour mission.

By substituting the given data values into Equation (3.30) we get:

$$R_{Sd}(100) = \sum_{j=0}^{1} \left[\left[(0.0005)(100) \right]^j e^{-(0.0005)(100)} \right] / j!$$
$$= 0.9988$$

Thus, the standby medical system reliability is 0.9988.

3.5 RELIABILITY ANALYSIS METHODS

Over the years many methods have been developed to perform various types of reliability analysis. This section presents three of these methods considered useful to perform reliability-related analysis in health care.

3.5.1 Markov Method

This is a powerful reliability analysis tool and is named for the Russian mathematician Andrei Andreyevich Markov (1856–1922). The method is widely used to analyze repairable systems with constant failure and repair rates. The following three assumptions are associated with the method[4]:

- The probability of occurrence from one system state to another in the finite time interval Δt is $\theta \Delta t$, where θ is the transition rate (i.e., constant failure or repair rate) from one system state to another.
- The probability of more than one transition occurrence in time interval Δt from one system state to another is negligible (i.e., $(\theta \Delta t)(\theta \Delta t) \rightarrow 0$).
- All occurrences are independent of each other.

The application of the method is demonstrated by solving the following example.

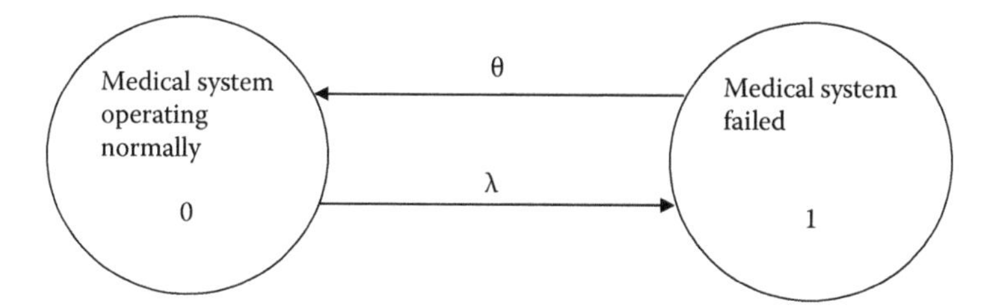

FIGURE 3.4 Medical system state space diagram.

Example 3.8

A medical system can either be in an operating or a failed state and its failure and repair rates, λ and θ, respectively, are constant. The system state space diagram is shown in Figure 3.4. The numerals in circles denote system states. Develop expressions for the medical system time dependent availability and unavailability by using the Markov approach.

Using the Markov method, we write down the following Equations for the Figure 3.4 diagram[4,5]:

$$P_0(t+\Delta t) = P_0(t)(1-\lambda\Delta t) + P_1(t)\theta\Delta t \tag{3.32}$$

$$P_1(t+\Delta t) = P_1(t)(1-\theta\Delta t) + P_0(t)\lambda\Delta t \tag{3.33}$$

where

λ is the medical system constant failure rate.
θ is the medical system constant repair rate.
$P_i(t)$ is the probability that the medical system is in state i at time t; for $i = 0, 1$.
$\lambda \Delta t$ is the probability of the medical system failure in finite time interval Δt.
$\theta \Delta t$ is the probability of the medical system repair in finite time interval Δt.
$(1 - \lambda \Delta t)$ is the probability of no failure in finite time interval Δt.
$(1 - \theta \Delta t)$ is the probability of no repair in finite time interval Δt.
$P_i(t + \Delta t)$ is the probability that the medical system is in state i at time $(t + \Delta t)$; for $i = 0, 1$.

In the limiting case, Equations (3.32) and (3.33) become:

$$\frac{dP_0(t)}{dt} + \lambda P_0(t) = P_1(t)\theta \tag{3.34}$$

$$\frac{dP_1(t)}{dt} + \theta P_1(t) = P_0(t)\lambda \tag{3.35}$$

At time $t = 0$, $P_0(0) = 1$ and $P_1(0) = 0$.

Solving Equations (3.34) and (3.35), we obtain:

$$P_0(t) = \frac{\theta}{(\lambda+\theta)} + \frac{\lambda}{(\lambda+\theta)} e^{-(\lambda+\theta)t} \tag{3.36}$$

$$P_1(t) = \frac{\lambda}{(\lambda+\theta)} - \frac{\lambda}{(\lambda+\theta)} e^{-(\lambda+\theta)t} \tag{3.37}$$

Thus, the system time dependent availability and unavailability, respectively, are:

$$AV_m(t) = P_0(t) = \frac{\theta}{(\lambda+\theta)} + \frac{\lambda}{(\lambda+\theta)} e^{-(\lambda+\theta)t} \tag{3.38}$$

and

$$UAV_m(t) = P_1(t) = \frac{\lambda}{(\lambda+\theta)} - \frac{\lambda}{(\lambda+\theta)} e^{-(\lambda+\theta)t} \tag{3.39}$$

where

$AV_m(t)$ is the medical system time dependent availability.
$UAV_m(t)$ is the medical system time dependent unavailability.

Thus, the medical system availability and unavailability are given by Equations (3.38) and (3.39), respectively.

3.5.2 Failure Modes and Effect Analysis

Failure modes and effect analysis (FMEA) is a bottom-up approach and it may simply be described as an effective method to perform analysis of each potential failure mode in the system to evaluate the effects or results of such failure modes on the system.[6] This method is called failure mode effects and criticality analysis (FMECA) when criticality analysis is added to it.[7-9] Criticality analysis is a quantitative approach for ranking critical failure mode effects by considering their probability of occurrence.

The history of FMEA may be traced back to the early 1950s with the development of flight control systems when the Untied States Navy's Bureau of Aeronautics, to establish an effective mechanism for reliability control over the detail design effort, developed a requirement called "failure analysis."[10]

FMEA can be performed by following the six steps presented below.[9,11]

Step 1: Define system and all its associated requirements.
Step 2: Establish ground rules to which the FMEA is to be performed.

Step 3: Describe the system and all its associated functional blocks.
Step 4: Identify failure modes and all their possible associated effects.
Step 5: Develop critical items list.
Step 6: Document the analysis.

All the above six steps are described in detail in ref. [5]. There are many advantages to performing FMEA. Some of these include a systematic approach to classify hardware failures; easy to comprehend; useful to compare designs; a useful tool to reduce engineering changes, development time, and cost; a visibility tool for managers; useful to improve communication among design interface personnel; useful to analyze small, large, and complex systems; identifies safety concerns to be focused on; a useful approach that begins from the detailed level and works upward; and improves customer satisfaction. A comprehensive list of publications on FMEA is available in ref. [12].

3.5.3 Fault Tree Analysis

Fault tree analysis (FTA) is a widely used method to determine the possible occurrence of undesirable events or failures in engineering systems. A fault tree may be described as a logical representation of the relationship of primary events that lead to a defined undesirable event known as the "top event" and is depicted using a tree structure with logic gates such as AND and OR. The method was developed in 1961 by W.H. Watson of Bell Telephone Laboratories to perform analysis of the Minuteman launch control system.[13]

FTA starts by identifying an undesirable event, called "top event," associated with a system. Fault events which could cause the top event are generated and connected by logic operators such as OR and AND. The OR gate provides a true (failure) output if one or more inputs are true (failures). Similarly, the AND gate provides a true (failure) output if all the inputs are true (failures). The fault tree construction proceeds by generation of fault events in a successive manner until the fault events need not be developed further. Also, in developing a fault tree, one successively asks the question: "How could this fault event occur?"

Although many symbols are used to construct fault trees, the four most commonly used symbols are shown in Figure 3.5.[14,15]

All the Figure 3.5 symbols are described below.

- **Circle.** It denotes a fault event or the failure of an elementary component. The failure parameters such as probability, failure and repair of a fault event are obtained from empirical data or other sources.
- **AND gate.** It denotes that an output fault event occurs if all the input fault events occur.
- **OR gate.** It denotes that an output fault event occurs if any one or more input fault events occur.

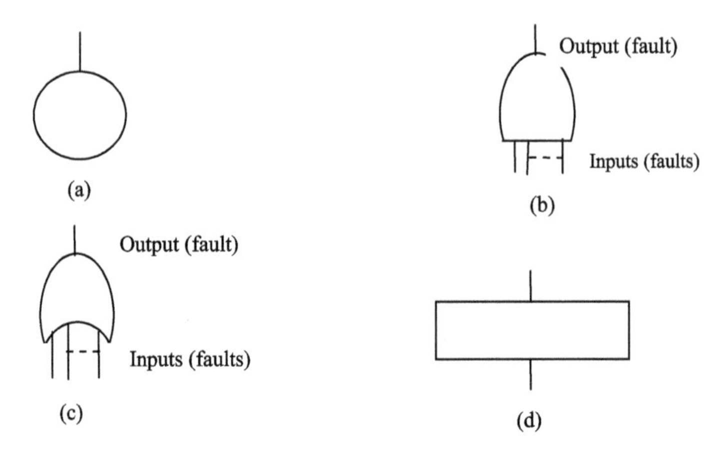

FIGURE 3.5 Most commonly used fault tree symbols: (**a**) circle, (**b**) AND gate, (**c**) OR gate, (**d**) rectangle.

- **Rectangle.** It denotes a fault event which results from the combination of fault events through the input of a logic gate such as AND and OR.

FTA can be performed by following the six steps shown in Figure 3.6.[14]

Example 3.9

A windowless surgery room has a switch and three light bulbs. The room can only be dark if the switch fails to close, all three light bulbs burn out, or there is no electricity. Develop a fault tree using the Figure 3.5 fault tree symbols for the undesired fault event "dark room" (i.e., room without light).

A fault tree for this example is shown in Figure 3.7.

The probability of AND and OR gates, shown in Figure 3.5, output events can be evaluated as follows[5,14]:

AND Gate

The probability of the occurrence of the AND gate output fault event, say X, is given by:

$$P(X)=\prod_{j=1}^{m} P(x_j) \tag{3.40}$$

where

m is the number of AND gate independent input fault events.
x_j is the AND gate input fault event j; for $j = 1, 2, 3, \ldots., m$.
$P(x_j)$ is the probability of occurrence of event x_j.

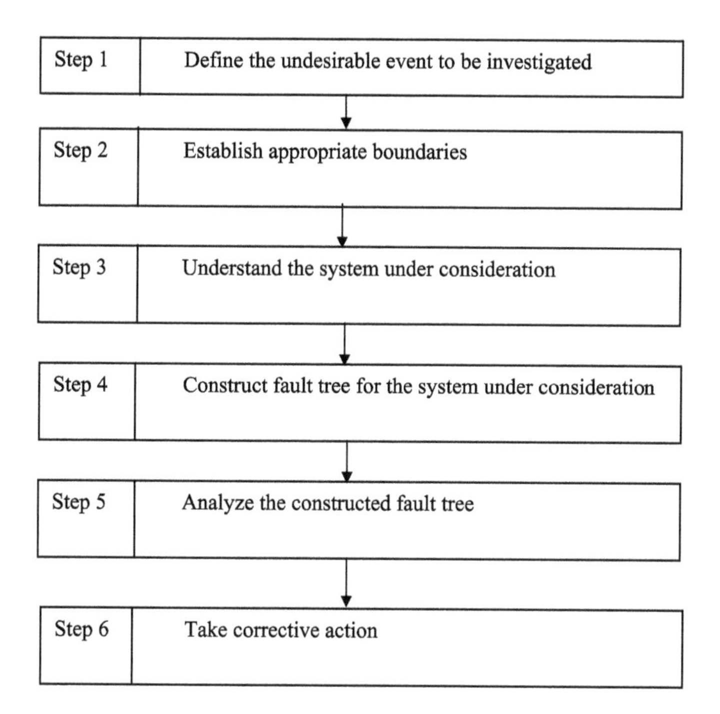

FIGURE 3.6 The steps for performing fault tree analysis.

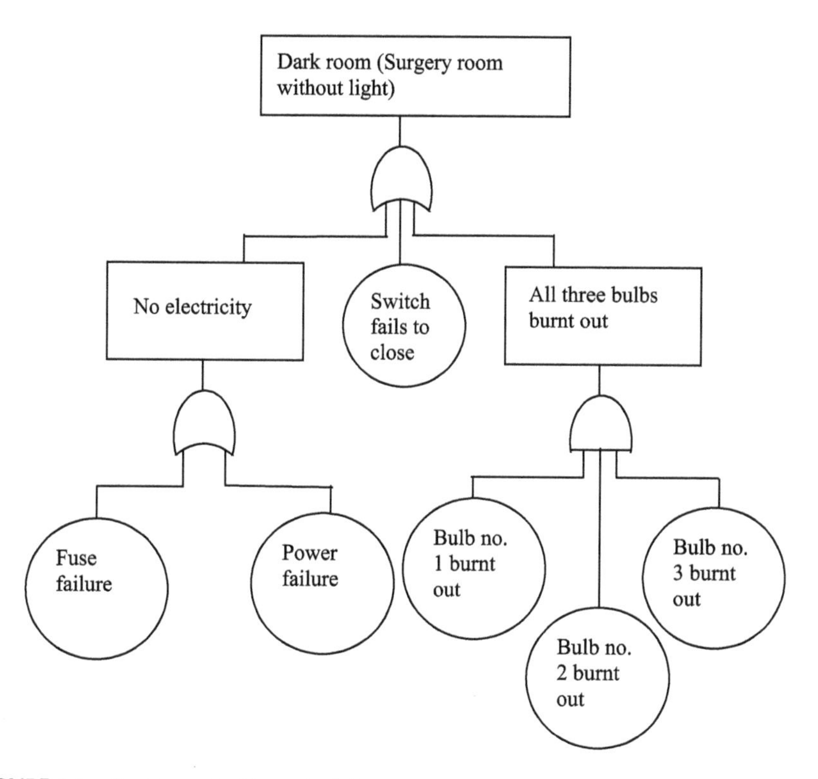

FIGURE 3.7 Fault tree for Example 3.9.

OR Gate

The probability of the occurrence of the OR gate output fault event, say Y, is given by:

$$P(Y)=1-\prod_{j=1}^{k}\{1-P(y_j)\} \tag{3.41}$$

where

k is the total number of OR gate independent input fault events.
y_j is the OR gate input fault event j; for $j = 1, 2, 3, \ldots, k$.
$P(y_j)$ is the probability of occurrence of event y_j.

3.6 HUMAN ERROR FACTS, FIGURES, AND CAUSES

Some facts and figures concerning human error are as follows[5,14]:

- Up to 90% of accidents both generally and in medical equipment/devices are due to human error.[15,16]
- Over 90% of the documented air traffic control system errors are caused by operators.[17]
- A study of 135 vessel failures during the period 1926 to 1988 revealed that 24.5% of the failures were caused by humans.[18]
- A study of 23,000 defects in the production of nuclear parts reported that around 82% of the defects were caused by human mistakes.[19,20]
- Over 50% of all technical medical equipment-related problems are caused by operators.[21]

There are many causes for the occurrence of human error. Some of these are poor equipment design, complex tasks, poor training or skill of concerned personnel, poor work layout, poorly written equipment operating and maintenance procedures, poor job environment, inadequate work tools, and poor motivation of involved personnel.[20,22]

3.7 HUMAN ERROR CLASSIFICATIONS

Humans make various types of errors. They may be categorized under the following classifications[20,22,23]:

- **Assembly errors.** This type of error occurs during product assembly due to humans. Some of the causes for the occurrence of assembly errors could be poor illumination, poorly designed work layout, poor communication of related information, poor blueprints and other related material, and excessive temperature in the work area.
- **Design errors.** These errors occur due to poor equipment design. Some of their causes are failure to implement human needs in the design, failure to

ensure the man-machine interaction effectiveness, and assigning inappropriate functions to humans.

- **Handling errors.** This type of error occurs basically due to poor transportation or storage facilities.
- **Inspection errors.** These errors occur because of less than 100% inspection accuracy. One typical example of an inspection error is rejecting and accepting in-tolerance and out-of-tolerance parts, respectively.
- **Installation errors.** These errors occur due to reasons such as failure to install equipment according to the manufacturer's specification and using the incorrect installation blueprints or instructions.
- **Maintenance errors.** These errors are due to oversights by the maintenance personnel. Two examples of maintenance errors are repairing the failed equipment incorrectly and calibrating equipment incorrectly.
- **Operator errors.** These errors are due to operator mistakes. Some of the causes for their occurrence are complex tasks, operator carelessness, poor environment, poor training, and lack of proper procedures.

3.8 HUMAN ERROR ANALYSIS METHODS

Over the years, many methods have been used to perform various types of human error analysis. These include failure modes and effect analysis (FMEA), fault tree analysis (FTA), Markov modeling, root cause analysis (RCA), and man-machine systems analysis (MMSA).[20,24] The first three (FMEA, FTA, and Markov modeling) are described in Section 3.5 and the remaining two (RCA and MMSA) are described below, separately.

3.8.1 Root Cause Analysis

Root cause analysis may be described as a systematic investigation method that uses information collected during assessing an accident to determine the underlying causes of the deficiencies that led to the accident.[25] This method was developed by the United States Department of Energy to investigate industrial incidents and is widely used in the area of health care. The following general steps are used to perform RCA in health care[26]:

Step 1: Educate all involved individuals about RCA.
Step 2: Inform all involved staff members as soon as the occurrence of a sentinel event is reported.
Step 3: Form an RCA team.
Step 4: Prepare for and hold the first team meeting.
Step 5: Determine the event sequence.
Step 6: Separate and highlight each event sequence that may have been a contributory cause in the sentinel event occurrence.
Step 7: Brainstorm about the possible factors surrounding the identified events that may have been contributory to the sentinel event occurrence.
Step 8: Affinitize the brainstorm session results.

Step 9: Develop an appropriate action plan.
Step 10: Distribute the RCA document and the action plan to all individuals involved.

Additional information on RCA is available in ref. [24].

3.8.2 Man–Machine Systems Analysis

MMSA was developed in the early 1950s for reducing human error-caused unwanted effects to an acceptable level in a system under consideration.[27] The following steps are associated with the method[24,27]:

Step 1: Define all appropriate functions and goals.
Step 2: Define the situational characteristics (e.g., air quality, illumination, etc., under which humans have to perform their tasks).
Step 3: Define characteristics of all involved individuals.
Step 4: Define the tasks performed by the individuals involved.
Step 5: Analyze tasks to identify potential error-likely conditions and other related difficulties.
Step 6: Estimate probability of occurrence of each potential human error.
Step 7: Estimate the chances that each potential human error will not be undetected or rectified.
Step 8: Determine all possible consequences if potential human errors remain undetected.
Step 9: Recommend changes.
Step 10: Re-evaluate all changes by repeating most of the above steps.

3.9 HUMAN ERROR DATA SOURCES

There are many sources for obtaining human error-related data. Some of these are as follows:

- Williams, H.I. Reliability Evaluation of the Human Component in Man-Machine Systems. *Electrical Manufacturing* 4 (1958): 78-82.
- American Institute for Research. An Index of Electronic Equipment Operability: Data Store.[28]
- Recht, J.L. Systems Safety Analysis: Error Rates and Costs. *National Safety News* February (1966): 20-23.
- Aviation Safety Reporting Program.[29]
- Nuclear Plant Reliability Data System.[30]
- Safety-Related Operator Action (SROA) Program.[31]
- Operational Performance Recording and Evaluation Data System.[32]
- Aerojet-General method.[33]
- Joos, D.W., Sabri, Z.A., Husseiny, A.A. Analysis of Gross Error Rates in the Operation of Commercial Nuclear Power Stations. *Nuclear Engineering and Design* 52 (1979): 265-300.

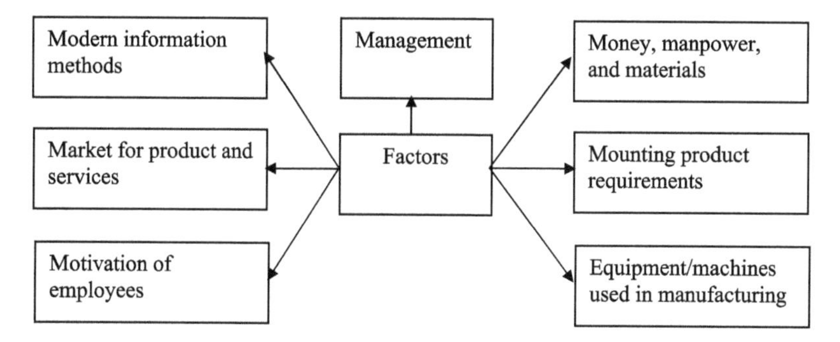

FIGURE 3.8 Factors directly influencing product quality and quality of services.

Some of the above data sources are described in detail in ref. [31]. Additional information on human error data sources is available in refs. [20, 34].

3.10 FACTORS AFFECTING PRODUCT QUALITY AND SERVICES

Many factors directly influence the quality of products and services. The important ones are shown in Figure 3.8.[35]

3.11 QUALITY ASSURANCE SYSTEM TASKS

The main goal of a quality assurance system is to effectively maintain the specified level of quality. Its important tasks are as follows[36]:

- Manage effectively the total quality assurance system
- Perform special quality-related studies
- Evaluate and control product quality in field environment
- Feedback appropriate quality-related information to management
- Consider with care the quality and reliability needs during the product development process
- Evaluate, plan, and control product quality effectively
- Develop personnel
- Monitor supplier quality assurance

3.12 TOTAL QUALITY MANAGEMENT VERSUS TRADITIONAL QUALITY ASSURANCE

Total quality management (TQM) is a widely used term and it was coined by Nancy Warren, a behavioral scientist, in 1985.[37] Each of its three words: total, quality, and management are described below, separately.

- **Total.** This basically calls for team effort of all concerned parties to satisfy customers. There are numerous factors that play an instrumental role in establishing an effective supplier–customer relationship, including customers making suppliers understand their obligations or needs, development of

Table 3.2 Comparisons Between Total Quality Management (TQM) and Traditional Quality Assurance Management (TQAM)

Area	TQM	TQAM
Objective	Eliminate the occurrence of errors	Find errors
Definition	Customer-driven	Product-driven
Customer	A well-defined approach to understand and meet customer requirements	Ambiguous understanding of consumer or customer needs
Cost	Better quality reduces cost and improves productivity	Improvements in quality lead to higher cost
Quality defined	Products suitable for consumer use	Products meet specifications
Quality responsibility	Everyone in the organization involved	Quality control inspection center/group

customer–supplier relationships on the basis of mutual trust and respect, and monitoring of suppliers' products and processes by customers on a regular basis.

- **Quality.** There are many definitions of quality. Nonetheless, quality must be viewed from the perspective of customers.
- **Management.** An effective approach to management is necessary to determine company ability to attain corporate goals and to allocate resources effectively.

To practice the TQM concept effectively, it is important to clearly understand the basic differences between TQM and traditional quality assurance management (TQAM). Table 3.2 presents comparisons between TQM and TQAM in six distinct areas.[38,39]

3.13 TQM ELEMENTS AND OBSTACLES TO TQM IMPLEMENTATION

There are many elements of TQM. The important ones are listed below.[40]

- Team work
- Management commitment and leadership
- Customer service
- Supplier participation
- Training
- Quality cost
- Statistical methods

Additional information on the above elements is available in ref. [40].

Over the years, people involved with the implementation of TQM have experienced various types of obstacles. A clear understanding of these obstacles is considered essential, prior to the initiation of the TQM implementation process. Some of these obstacles in the form of questions are as follows[41]:

- Who will establish the TQM vision?
- Will upper management support the introduction of the TQM program whole-heartedly?
- What are the ways and means to convince involved people that TQM is different?
- Do the management personnel understand the TQM objective?
- What are the ways and means to convince people of the need to change?
- Is there sufficient time to implement the TQM program effectively?
- Is it possible to obtain adequate support of managers and their concerned subordinates possessing an "independent" attitude?
- Is it possible to quantify all customer requirements? If so, how?

3.14 PROBLEMS

1. What is the bathtub hazard rate curve? Discuss it in detail.
2. Prove Equation (3.1) by using Equation (3.6).
3. Assume that the hazard rate of a piece of medical equipment is expressed by:

$$\lambda(t)=\frac{\beta}{\theta}\left(\frac{t}{\theta}\right)^{\beta-1} \tag{3.42}$$

where

t is time.
θ is the scale parameter.
β is the shape parameter.

Obtain an expression for the medical equipment reliability.

4. Prove by using Equation (3.3) that Equations (3.8) and (3.10) yield identical results.
5. Assume that a medical system is composed of three active, independent, and identical units and at least one of the units must operate normally for system success. The unit failure rate is 0.0005 failures per hour. Calculate the medical system reliability for a 200-hour mission and mean time to failure.
6. What is the difference between parallel configuration and standby system?
7. Prove that the sum of Equations (3.36) and (3.37) is equal to unity.
8. Describe RCA.
9. Discuss the following types of errors:
   - Maintenance errors
   - Design errors
   - Operator errors
   - Inspection errors
10. Compare TQM with TQAM.
11. List at least seven important tasks of a quality assurance system.

REFERENCES

1. Shapero, A., Cooper, J.I., Rappaport, M., Shaeffer, K.H., Bates, C.J. Human Engineering Testing and Malfunction Data Collection in Weapon System Programs. WADD Technical Report, 60-36. Dayton, OH: Wright-Patterson Air Force Base, February 1960.
2. Feigenbaum, A.V. *Total Quality Control.* New York: McGraw-Hill, 1983.
3. Kapur, K.C. Reliability and Maintainability. In *Handbook of Industrial Engineering,* edited by G. Salvendy, 8.5.1-8.5.34. New York: John Wiley & Sons, 1982.
4. Shooman, M.L. *Probabilistic Reliability: An Engineering Approach.* New York: McGraw-Hill, 1968.
5. Dhillon, B.S. *Design Reliability: Fundamentals and Applications.* Boca Raton, FL: CRC Press, 1999.
6. Omdahl, T.P., ed. *Reliability, Availability and Maintainability (RAM) Dictionary.* Milwaukee, WI: American Society for Quality Control (ASQC) Press, 1988.
7. Procedures for Performing a Failure Mode, Effects, and Criticality Analysis. MIL-STD-1629. Washington, DC: Department of Defense, 1980.
8. Grant Ireson, W., Coombs, C.F., Moss, R.Y. *Handbook of Reliability Engineering and Management.* New York: McGraw-Hill, 1996.
9. Dhillon, B.S. *Systems Reliability, Maintainability, and Management.* New York: Petrocelli Books, 1983.
10. General Specification for Design, Installation, and Test of Aircraft Flight Control Systems. MIL-F-18372 (Aer); Para 3.5.2.3. Washington, DC: Bureau of Naval Weapons, Department of the Navy.
11. Jordan, W.E. Failure Modes, Effects, and Criticality Analyses. Proceedings of the Annual Reliability and Maintainability Symposium, 1971, 30-37.
12. Dhillon, B.S. Failure Mode and Effects Analysis: Bibliography. *Microelectronics and Reliability* 32 (1992): 719-731.
13. Fussell, J.E., Powers, G.J., Bennets, R.G. Fault Trees: A State of the Art Discussion. *IEEE Transactions on Reliability* 30 (1974): 51-55.
14. Dhillon, B.S. *Medical Device Reliability and Associated Areas.* Boca Raton, FL: CRC Press, 2000.
15. Nobel, J.L. Medical Device Failures and Adverse Effects. *Pediatric Emergency Care* 7 (1991): 120-123.
16. Askern, W.B., Regulinski, T.L. Quantifying Human Performance for Reliability Analysis of Systems. *Human Factors* 11 (1969): 393-396.
17. Kenney, G.C., Spahn, M.J., Amato, R.A. The Human Element in Air Traffic Control: Observations and Analysis of Performance of Controllers and Supervisors in Providing Air Traffic Control Separation Services. Report MTR-7655. Los Angeles: METREK Division, MITRE Corporation, December 1977.
18. Organizational Management and Human Factors in Quantitative Risk Assessment. Report 33/192 (Report 1). London: British Health and Safety Executive, 1992.
19. Rook, L.W. Reduction of Human Error in Industrial Production. Report SCTM 93-63 (14). Albuquerque, NM: Sandia Laboratories, June 1962.
20. Dhillon, B.S. *Human Reliability: With Human Factors.* New York: Pergamon Press, 1986.
21. Dhillon, B.S. Reliability Technology in Health Care Systems. Proceedings of the IASTED International Symposium on Computers and Advanced Technology in Medicine, Healthcare, and Bioengineering, 1990, 84-87.
22. Meister, D. The Problem of Human-Initiated Failures. Proceedings of the 8th National Symposium on Reliability and Quality Control, 1962, 234-239.

23. Cooper, J.I. Human-Initiated Failures and Man-Function Reporting. *IRE Transactions on Human Factors* 10 (1961): 104-109.
24. Dhillon, B.S. *Human Reliability and Error in Medical System*. River Edge, NJ: World Scientific Publishing, 2003.
25. Latino, R.J. Automating Root Cause Analysis. In *Error Reduction in Healthcare*, edited by P.L. Spath, 155-164. New York: John Wiley & Sons, 2000.
26. Burke, A. Root Cause Analysis. Report, 2002. Available from the Wild Iris Medical Education, P.O. Box 257, Comptche, CA 95427.
27. Miller, R.B. A Method for Man-Machine Task Analysis. Report 53-137. US Air Force (USAF), Ohio: Wright Air Development Center, Wright-Patterson Air Force Base, 1953.
28. Munger, S.J., Smith, R.W., Pyne, D. An Index of Electronic Equipment Operability: Data Store. Report C 43-1/62 RP (1). Pittsburgh, PA: American Institute for Research, 1962.
29. Aviation Safety Reporting Program. FAA Advisory Circular 00-46B. Washington, DC: Federal Aviation Administration, June 15, 1979.
30. Reporting Procedures Manual for the Nuclear Plant Reliability Data System (NPRDS). San Antonio, TX: South-West Research Institute, December 1980.
31. Tomiller, D.A., Eckel, J.S., Kozinsky, E.J. Human Reliability Data Bank for Nuclear Power Plant Operations: A Review of Existing Human Reliability Data Banks. Report NUREG/CR2744/1. Washington, DC: U.S. Nuclear Regulatory Commission, 1982.
32. Urmston, R. Operational Performance Recording and Evaluation Data System (OPREDS). Descriptive Brochures, Code 3400. San Diego, CA: Navy Electronics Laboratory Center, November 1971.
33. Irwin, I.A., Levitz, J.J., Freed, A.M. Human Reliability in the Performance of Maintenance. Report LRP 317/TDR-63-218. Sacramento, CA: Aerojet-General Corporation, 1964.
34. Dhillon, B.S. Human Error Data Banks. *Microelectronics and Reliability* 30 (1990): 963-971.
35. Feigenbaum, A.V. *Total Quality Control*. New York: McGraw-Hill, 1983.
36. The Quality World of Allis-Chalmers. *Quality Assurance* 9 (1970): 13-17.
37. Walton, M. *Deming Management at Work*. New York: Putnam, 1990.
38. Schmidt, W.H., Finnegan, J.P. *The Race Without a Finish Line: America's Quest for Total Quality*. San Francisco: Jossey-Bass Publishers, 1992.
39. Madu, C.N., Chu-hua, K. Strategic Total Quality Management (STQM). In *Management of New Technologies for Global Competitiveness*, edited by C.N. Madu, 3-25. Westport, CT: Quorum Books, 1993.
40. Burati, J.L., Matthews, M.F., Kalidindi, S.N. Quality Management Organization and Techniques. *Journal of Construction Engineering and Management* 118 (1992): 112-128.
41. Klein, R.A. Achieve Total Quality Management. *Chemical Engineering Progress* November (1991): 83-86.

4 Medical Device Safety and Quality Assurance

4.1 INTRODUCTION

Safety is an important consideration in medical devices as a medical device must not only be reliable but also safe. The history of safety with respect to health may be traced back to circa 2000 BC in what is known as the "Code of Hammurabi"; this code contains clauses concerning injuries and financial damages against those causing injury to others.[1-3]

In modern times, the passage of the Occupational Safety and Health Act (OSHA) in 1970 is regarded as a major milestone with regard to health and safety in the United States. The other two major milestones, particularly in regard to medical devices, in the United States are the Medical Device Amendments of 1976 and the Safe Medical Device Act (SMDA) of 1990.

The history of quality assurance may also be traced back to ancient times; for example, around 1450 BC, some Egyptian wall paintings show evidence of inspection-related activities.[4] However, in modern times, the Western Electric Company was the first organization in the United States to set up an inspection department in the early 1900s.[5] In 1976, amendments to the Federal Food, Drug, and Cosmetic Act concerning medical devices helped to establish a complex statutory framework to allow the U.S. Food and Drug Administration (FDA) to regulate most aspects of medical devices, from testing to marketing.

This chapter presents various important aspects of medical device safety and quality assurance.

4.2 MEDICAL DEVICE SAFETY VERSUS RELIABILITY AND TYPES OF MEDICAL DEVICE SAFETY

Although both safety and reliability are good things to which all medical devices should aspire, they are distinct concepts and at times can have conflicting concerns.[6]

A safe medical system may be described in the simplest terms as a system that does not cause a high degree of risk to property, equipment, or people. Risk is an undesirable event that can occur and is measured in terms of probability and severity. More specifically, medical device safety is simply a concern with malfunctions or failures that introduce hazards and is expressed with respect to the level of risk, but not in terms of satisfying specified requirements. On the other hand, the reliability of a medical device/equipment is one minus the probability of failure to satisfy its specified requirements.

A medical device is still considered safe even when it fails frequently but without mishaps. On the other hand, if a device functions normally all the time but regularly puts people at risk, then such a device is considered reliable but unsafe. Two examples of safe/unreliable and reliable/unsafe medical devices, respectively, are as follows[6]:

Example I: A pacemaker that does not always operate properly at the programmed rate for a small number of patients is not a safety concern. In this case, the pacemaker is safe but unreliable.

Example II: A pacemaker that can pace, says at 112 beats per minute in any situation, is considered highly reliable. However, if the patient is in cardiac failure, a high pacing rate, from the medical aspect, is considered quite inappropriate. In this case, the pacemaker is reliable but unsafe.

Medical device safety may be classified under the following three categories[7]:

- **Unconditional safety.** This type of safety is preferred over all other safety types because it is most effective. It requires elimination of all risks associated with medical devices through the design process. All in all, it is to be noted that the use of appropriate warnings complements satisfactory device design, but does not replace it.
- **Conditional safety.** This type of safety is used in situations when it is not feasible to realize unconditional safety. For example, in the case of an x-ray or a laser surgical device, it is impossible to avoid dangerous radiation emissions. However, it is quite possible to reduce risk through actions, such as including a locking switch that permits device activation by authorized personnel only or limiting access to therapy rooms.
- **Descriptive safety.** This type of safety with regard to transport, replacement, maintenance, mounting, operation, and connection may simply be various requirements, including "Handle with care," "This side up," and "Not for explosive zones."

4.3 MEDICAL DEVICE-RELATED SAFETY REQUIREMENTS AND SAFETY IN DEVICE LIFE CYCLE

Governments and other regulatory bodies have imposed various types of requirements on medical devices, directly or indirectly, with respect to safety. These requirements may be grouped under three categories, as shown in Figure 4.1.[7] These are safe design, sufficient information, and safe function. The "safe design" category is composed of the following seven elements:

- Protection against electrical shock
- Protection against radiation hazards
- Care for hygienic factors
- Excessive heating prevention
- Mechanical hazard prevention
- Care for environmental conditions
- Proper material choice with respect to chemical, biological, and mechanical factors

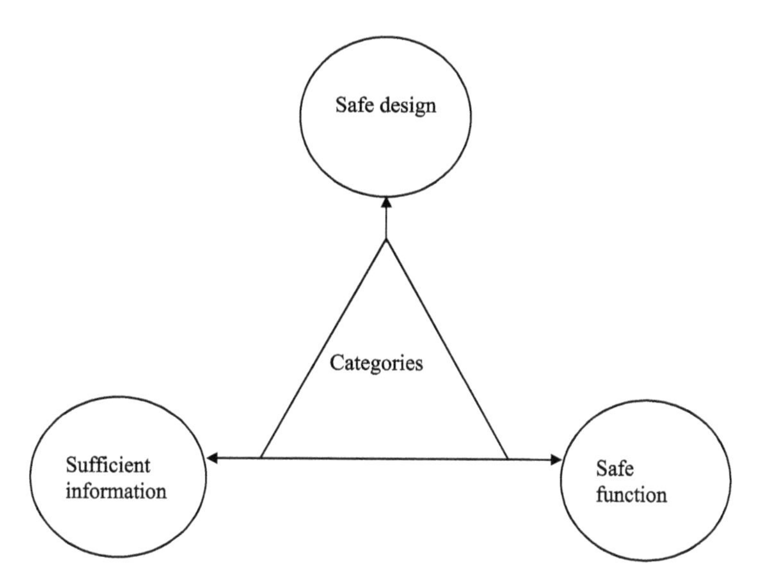

FIGURE 4.1 Categories of medical device safety-related requirements imposed by governments and other regulatory bodies.

Similarly, the "safe function" category has three elements: accuracy of measurements, reliability, and warning for or prevention of dangerous outputs. Finally, the "sufficient information" category elements include effective labelling, instructions for use, production, packaging, and accompanying documentation.

To have a safe medical device, it is essential to consider safety throughout its life cycle. The life cycle of a medical device may be divided into five phases, as shown in Figure 4.2.[3,8] These are concept phase, definition phase, development phase, production phase, and deployment phase.

In the concept phase future technical projections and past data and experiences are the basis for the medical device under consideration. Thus, during this phase, safety problems and their associated impacts are identified and evaluated by using tools such as the preliminary hazard analysis. At the end of this phase, some of the questions that can be asked in regard to the device safety are as follows:

- Are all the necessary basic safety-related design requirements for the phase in place, so that the definition phase can be started without any problem whatsoever?
- Are the hazards properly identified and evaluated for developing hazard controls?
- Is the risk analysis initiated to develop mechanisms for hazard controls?

The basic purpose of the definition phase is to provide verification of the device initial design and engineering. Thus, the preliminary hazard analysis of the concept phase is updated along with the initiation of subsystem hazard analysis and their subsequent integration into the device hazard analysis. In this phase, the application of methods such as fault tree analysis and fault hazard analysis could be quite useful in examin-

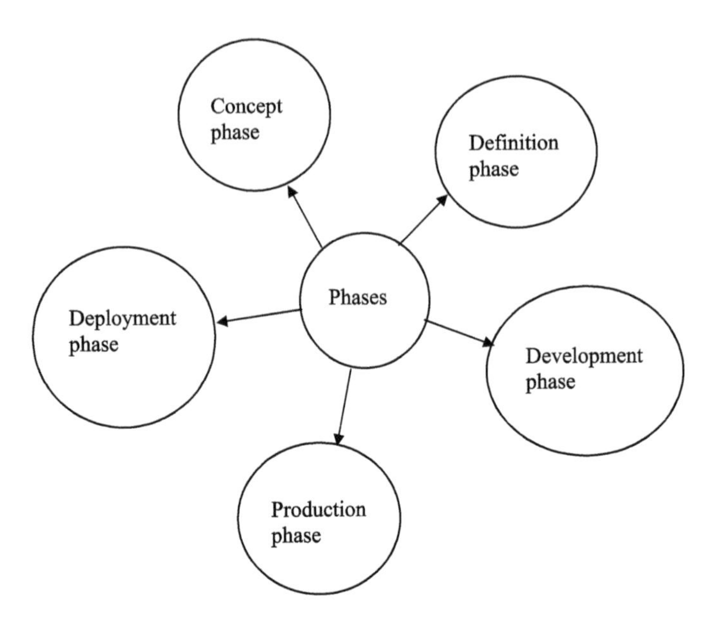

FIGURE 4.2 Medical device life cycle phases.

ing specific known hazards and their effects. Finally, it is added that the system definition will initially result in the acceptance of a desirable general design, even though due to the incompleteness of the design not all hazards will be clearly known.

In the device development phase, effort is directed on areas such as producibility engineering, operational use, environmental impact, and integrated logistics support. The preliminary hazard analysis is developed further because of more completeness of the device design, in addition to performing comprehensive operating hazard analysis, by utilizing prototype analysis and testing results, to evaluate man-machine hazards.

In the production phase, a device safety engineering report is prepared using the data gathered during the phase. The report identifies and documents all concerned hazards with the resulting device.

In the deployment phase, data relating to accidents, incidents, failures, etc., are collected, the safety analysis is updated, and changes to the device are reviewed by the safety professionals.

4.4 SAFETY ANALYSIS TOOLS FOR MEDICAL DEVICES AND CONSIDERATIONS FOR SELECTING SAFETY ANALYSIS METHODS

Over the years many methods and techniques have been developed to perform various types of safety analysis. Many of these methods can also be used to perform safety analysis of medical devices. Some of these methods are shown in Figure 4.3.[8-14] Three of these methods (fault tree analysis, failure modes and effect analysis, and hazard and operability review) are described in Chapters 3 and 9 and in refs. [3, 9,

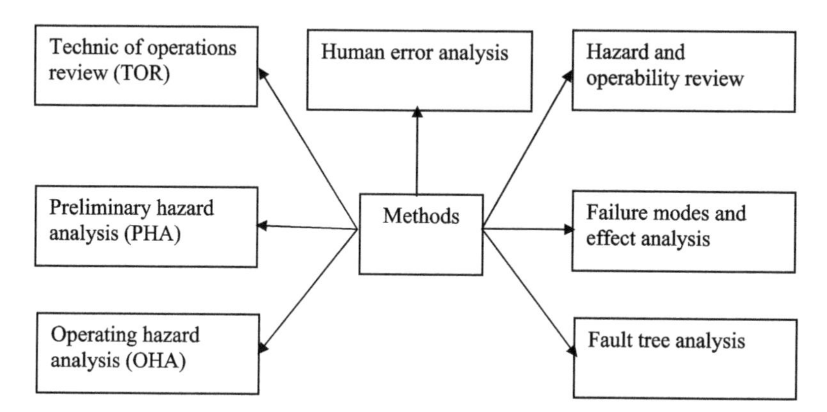

FIGURE 4.3 Methods for performing safety analysis of medical devices.

14, 15]. The others (preliminary hazard analysis, operating hazard analysis, technic of operations review, and human error analysis) are described below, separately.

Preliminary Hazard Analysis

Preliminary hazard analysis may be called a conceptual design approach, because it is the first hazard analysis performed on a new product. Some of the main objectives of performing PHA are as follows:

- To identify safety critical areas
- To identify safety design criteria to be used
- To evaluate hazards

Past experiences indicate that usually PHA is started during the primary design phase so that all appropriate safety aspects are included into trade-off studies. There are many PHA-associated activities. Some of these are as follows[3]:

- Reviewing past safety relevant experiences
- Examining compliance with safety requirements and other regulations concerning items, such as toxic substances, personnel safety, and environmental hazards
- Listing primary energy sources
- Evaluating energy sources to determine the provisions established for their control

For effective hazard identification, the areas of item/device/system design to be considered by PHA are as follows[3,8]:

- **Normal and abnormal environmental problems.** These problems include extreme temperatures, noise, electrostatic discharge, shock, x-ray radiation, and laser radiation.

- **Operating test/maintenance/procedural problems.** These problems include human error and emergency requirements, such as rescue, egress, or survival.
- **Interface safety-related problems.** These problems include electromagnetic interference, material compatibilities, and inadvertent activation.
- **Facilities and support equipment with commensurate training.** For effective usage these two items should be examined with respect to factors such as provisions for storage, assembly, and testing hazardous substances.
- **Energy source hazardous components.** Two typical examples of such components are propellants and pressure systems.

Worksheets for performing preliminary hazards analysis are available in refs. [8, 12].

Operating Hazard Analysis (OHA)

This method focuses on hazards resulting from activities/tasks for operating system functions that happen as the system or device is stored/transported/used. Usually, this type of analysis is initiated early in the device/system development cycle so that appropriate inputs to technical orders are provided effectively, which in turn govern device/system testing. Professionals involved in the performance of OHA need appropriate engineering descriptions of the device/system with available support facilities. Furthermore, this type of analysis is performed using a form that requires information on items such as details of the operational event, hazard description and effects, hazard control, and all associated requirements.

The application of the OHA will provide a basis for safety considerations such as follows[3,8,11]:

- Design modifications to eradicate hazards
- To provide appropriate safety devices and safety guards
- Identification of device/item functions relating to hazardous occurrences
- Special safety-related procedures in regard to handling, transporting, storing, servicing, and training
- Development of operation-related warning, emergency procedures, or special instructions

Technic of Operations Review (TOR)

This method allows both management and workers to work together in performing analysis of workplace-related incidents, accidents, and failures. The method was developed by D.A. Weaver in the early 1970s and uses a worksheet containing simple and straightforward terms requiring yes/no decisions.[10] The following eight steps are associated with the method[14]:

Step 1: Form the TOR team.
Step 2: Hold a roundtable session to impart common knowledge to all team members.
Step 3: Identify through consensus one important factor that caused the incident/accidents.
Step 4: Use the team consensus in responding to a sequence of yes/no options.
Step 5: Evaluate the identified factors through team consensus.
Step 6: Prioritize the contributing factors starting with the most critical.
Step 7: Develop preventive/corrective strategies with respect to each contributing factor.
Step 8: Prepare the final document and implement strategies.

Finally, it is to be noted that the strength of the method stems from the involvement of line personnel in the analysis and its weakness being an after-the-fact process.

Human Error Analysis (HEA)

This is another safety analysis approach that can be used to highlight various types of hazards prior to their occurrence in the form of accidents. Past experiences indicate that there could be two distinct and effective approaches to HEA:

- Actually performing tasks, in order to get first hand information on hazards
- Observing individuals during their work period in regard to hazards

All in all, it is recommended to perform HEA in conjunction with failure modes and effect analysis (FMEA) and hazard and operability review (HAZOP) to obtain best results. Both FMEA and HAZOP are described in Chapters 3 and 9 and in refs. [3, 9, 14, 15].

4.4.1 Considerations for Selecting Safety Analysis Methods

For effective applications of safety analysis methods, a careful consideration is necessary in their selection and implementation. Thus, prior to their selection and implementation for a given situation, relevant questions on areas such as follows should be asked[12]:

- Time when the results are required
- Ways and means for acquiring information from subcontractors (if applicable)
- Format and the degree of detail of the end result data for recipients
- Organizations/individuals who will be using the end results
- Type of information, data, drawings, etc. required prior to performing the study under consideration
- Time frame for the initiation of analysis and its review, update, submission, and completion

4.5 REGULATORY COMPLIANCE OF MEDICAL DEVICE QUALITY ASSURANCE AND PROGRAM FOR ASSURING MEDICAL DEVICE DESIGN QUALITY

In order to produce good quality medical devices, the FDA and the ISO (International Organization for Standardization) are actively playing an instrumental role through their good manufacturing practices (GMP) regulation and ISO 9000 family of quality system standards, respectively. Both these organizations require manufacturers to follow a comprehensive approach to medical device quality.

A procedure, in the form of a quality assurance manual, to assist medical device manufacturers to satisfy the GMP regulation and applicable ISO 9000 requirements can be divided into three areas[16]:

Area I is concerned with items such as outlining the organization/company policy with respect to the manufacture of medical devices, the quality assurance department's authority and responsibility to implement the policy, and defining the type of records to be maintained.

Area II is concerned with defining the company policy with respect to the administration of the quality assurance department and its subdivisions, such as receiving inspection and tools and gauge inspection, outlining the policy for internal audits and product qualification, and outlining the organizational chart.

Area III is concerned with outlining quality assurance directives to implement and monitor device conformance and procedural compliance according to the GMP regulation and the applicable ISO requirements. These directives cover the following sub-areas[16]:

- Sterilization process and control
- Procedure for FDA inspection
- Quality audits
- Equipment
- Design control
- Facility control for the manufacturing of medical devices
- Failure analysis
- Statistical quality control and sampling
- Field corrective action
- Component/warehouse control
- Lot numbering and traceability
- Approval and control of labels, labeling, and advertisement
- Complaint report procedure
- Personnel
- Supplier and subcontractor quality audits
- Control of inspection stamps
- Control of measuring equipment and tooling
- In-process and final inspection
- Receiving inspection

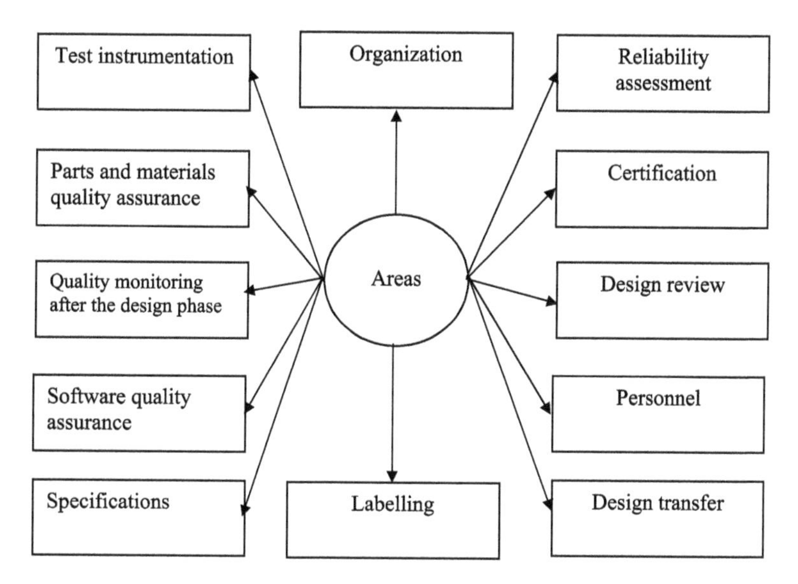

FIGURE 4.4 Areas of the FDA design quality assurance program.

Design phase is an important phase of engineering products including medical devices. More specifically, it is the phase during which the devices' inherent reliability, safety, and effectiveness are established. The existence of an effective quality assurance program during this phase is essential. The FDA has developed a document entitled "Pre-production Quality Assurance Planning: Recommendations for Medical Device Manufacturers" to assist manufacturers in planning and implementing their preproduction quality assurance programs.[17] The preproduction or design quality assurance program recommended by the FDA is divided into 12 areas, as shown in Figure 4.4.[17] All these are described below, separately.

Organization

This area is concerned with various organizational aspects of the program including the organizational elements and authorities necessary to develop the program, to execute program requirements, formal establishment of audit program, and formal documentation of the specified program goals.

Specifications

After establishing characteristics such as performance and physical for the medical device under consideration, they should be translated into design specifications through which the suitable design can be developed, evaluated, and controlled. These specifications should address issues such as reliability, safety, stability, and precision. The specification document should be reviewed by professionals belonging to areas such as quality assurance, reliability, research and development, and marketing.

All in all, two factors that must be carefully considered in the specification document are design changes and system compatibility. The design changes are those changes made to the specification during the research and development phase that are formally accepted as the design changes. The system compatibility is concerned with the compatibility of the device with other devices in the intended operating system, to assure satisfactory functioning of the total system. For example, ensure to incorporate disposable electrodes with cardiac monitors and breathing circuits with ventilators.

Parts and Materials Quality Assurance

This area is concerned with assuring that parts and materials used in device designs have adequate reliability levels to achieve their specified goals effectively. This requires establishment and implementation of effective parts and materials quality assurance programs by all concerned medical device manufacturers. These programs should cover areas such as qualification and ongoing verification of parts and materials quality, selection, and whether fabricated in-house or purchased from vendors.

Design Transfer

After the translation of the design into physical entity and then verifying its technical adequacy through testing under the simulated or real-life use environments, the design should be approved. A careful consideration is required when moving from laboratory to scaled-up production, as standards/methods/procedures may not be transferred effectively or it is quite possible that additional manufacturing processes are added.

Reliability Assessment

Reliability assessment is the process of prediction and demonstration used to estimate the basic reliability of a medical device and it should be performed for new and modified device designs. The appropriateness and extent of reliability assessment is determined by the degree of risk the device under consideration presents to its users.

The reliability assessment process can be started by theoretical/statistical approaches by first evaluating reliability of each component and then of the total device/system. Although this approach provides useful estimates of reliability, for better reliability estimates, the device/system should be tested under simulated or real-life use environment.

Design Review

The objective of design review is to highlight and eradicate design-related deficiencies as soon as possible because their implementation will be less costly. The design review program should be documented in adequate detail and it should include items such as organizational units involved, schedule, a checklist of variables to be evaluated, procedures used, and process flow diagrams.

Although the frequency and extent of design reviews will depend on various factors including the complexity and significance of the device/system, the assessment should include items such as subsystems, components, support documentation, packaging and labeling, and software (if applicable). During the formation of the design review team, careful consideration must be given to factors such as the experience and the background of the team members. More specifically, the team members should be from areas such as engineering, research and development, manufacturing, quality assurance, marketing, and servicing.

Personnel

This area calls for the performance of design-related activities, including design review, analysis, and testing by individuals having adequate experience and training.

Software Quality Assurance

The establishment of a software quality assurance program is essential when a medical device design incorporates software developed in-house. The main goal of this program should be software reliability, maintainability, testability, and correctness. All in all, the program should include a protocol for formal review and validation.

Test Instrumentation

Test instrumentation is concerned with calibrating and maintaining all pieces of equipment used in the qualification of the device design under consideration. In order to achieve effective end results, such equipment should be kept under a formal calibration and maintenance program.

Labeling

This area incorporates items such as display labels, manuals, charts, inserts, panels, and recommended test and calibration protocols. The design review process should also review labeling with care to assure that it contains easy to understand instructions or directions for end users as well as that it complies fully in regard to all applicable laws and regulations.

The device qualification testing should also incorporate the labeling aspect as well.

Quality Monitoring After the Design Phase

The effort to produce reliable, safe, and effective medical devices is a continuous process throughout their life cycle. More specifically, it does not end at the completion of the design phase but it continues during manufacturing and field use phases as well. Therefore, medical device manufacturers should establish a program for purposes such as to analyze quality problems, to take corrective actions to prevent recurrence of identified problems, and to identify failure patterns.

Certification

Subsequent to the successful passing of preproduction qualification testing by the initial production units, it is quite essential to perform a formal technical review with the aim of assuring the adequacy of the design, production, and quality assurance procedures. The review should include the determination of factors such as listed below.

- Adequacy of specifications and the specification change control program
- Effectiveness of test methods used to evaluate compliance with regard to applicably approved specifications
- Adequacy of the overall quality assurance plan
- Resolution of any discrepancy between the actual end device and the final approved device specifications

4.6 METHODS FOR MEDICAL DEVICE QUALITY ASSURANCE

Over the years, many methods have been developed for use in quality work.[18] Some of these methods considered useful to improve quality of medical devices are presented below.[14,18-21]

4.6.1 Flowcharts

Flowcharts are often used to describe processes in considerable detail by graphically showing the associated steps in proper order. A flowchart can be either simple or complex, composed of many symbols, boxes, etc. The complex version shows the process steps in the proper sequence and associated step conditions, as well as the related constraints using elements such as arrows, if/then statements, or yes/no choices.

All in all, a good flowchart shows all process steps under consideration or analysis by the quality improvement team members, suggests important areas for improvement, serves as an effective tool to solve or explain a problem, and identifies critical process points for control. Additional information on flowcharts is available in refs. [18, 20].

4.6.2 Check Sheets

Check sheets are often used to collect various types of data by operating personnel, so that such data can easily and efficiently be analyzed and used. In particular, they are quite useful to organize data by classification and indicate the number of times each data value occurs. This type of information becomes increasingly important and useful as more data are collected. Nonetheless, for the effectiveness of this approach, there must always be 50 or more observations available to be charted.

Some of the important points concerning the check sheets are listed below.

- Normally, the project team design check sheets and the check sheet forms are individualized for each scenario.
- Make the check sheet user-friendly as much as possible and include data on location and time.
- Creativity plays an important role during check sheet design.
- Check sheets are an effective tool for operators to spot problems, when the frequency of a specific defect and its occurrence frequency in a particular location are known.

Additional information on check sheets is available in refs. [18, 20].

4.6.3 Scatter Diagram

The scatter diagram is a simple and straightforward approach for determining how two given variables are related or if there is a cause-and-effect relationship between the two variables. In these diagrams, the horizontal axis represents the measurement values of one variable and the vertical axis, the measurements of the other variable. In situations when it is essential to fit a straight line to the plotted data points for obtaining a prediction equation, a line can be drawn on the scatter diagram either mathematically using the least square method or simply by sight.

Finally, it is added that the scatter diagram cannot prove that one variable causes the change in the other but only the existence of their relationship and its level of strength. Additional information on the scatter diagram is available in refs. [18, 20].

4.6.4 Histogram

A histogram plots data in a frequency distribution table and is used when good clarity is essential. Its main distinction from a check sheet is that its data values are grouped into rows to lose the identity of individual values. The histogram may be called the first "statistical" process control method because it can describe process variations.

The histogram shape depicts the nature of the distribution of the data values as well as central tendency and variability. Also, specification limits may be employed to depict process capability. All in all, a histogram can give sufficient information regarding a quality-related problem, thus providing a good basis to make decisions without performing any additional analysis. Additional information on histograms is available in refs. [18, 20].

4.6.5 Cause-and-Effect Diagram

This method was developed in 1943 by a Japanese quality expert named Kaoru Ishikawa (hence, also called the Ishikawa diagram); this cause-and-effect diagram is also known as the "fishbone diagram" because of its resemblance to the bones of a fish. The diagram depicts desirable or undesirable outcome as an effect and the causes leading to or potentially leading to the effect. Usually, the diagram is used to investigate a "bad" effect and to take necessary actions to rectify the associated causes.

For example, in the cause-and-effect diagram for the total quality management effort, the effect could be customer satisfaction and the associated major causes: manpower, machines, materials, and methods. A careful analysis of these four causes can serve as an effective tool in identifying possible quality problems and inspection points. A cause-and-effect diagram can be developed by following the five steps listed below.

Step 1: Establish problem statement.
Step 2: Brainstorm to highlight relevant possible causes.
Step 3: Group causes under major classifications.
Step 4: Develop the diagram.
Step 5: Refine cause classifications.

Some of the advantages of the cause-and-effect diagram are useful to generate ideas, useful to identify the root cause, and useful to guide further inquiry. The major disadvantage of the cause-and-effect diagram approach is that users can overlook critical and complex interactions between causes. Additional information on the cause-and-effect diagram is available in Chapter 11 and in refs. [18, 20].

4.6.6 Pareto Diagram

The Pareto diagram is named after Vilfredo Pareto (1848–1923), an Italian economist. Its application in quality work was popularized by Joseph Juran, a quality management guru, who believed that 80% of quality problems are the result of only 20% of the causes.

Basically, the Pareto diagram is a type of frequency chart that arranges data values in a hierarchical order; thus, it helps to highlight the most pressing problems to be eradicated first.

The following five steps are associated with the construction of a Pareto diagram[19]:

Step 1: Determinethe most suitable approach to classify data, i.e., by problem, cause, nonconformity, etc.
Step 2: Determine what to use to rank characteristics, i.e., dollars or frequency.
Step 3: Collect data for time intervals.
Step 4: Summarize the data and rank classifications from largest to smallest.
Step 5: Construct the Pareto diagram and then determine the significant few.

Although the Pareto method is quite useful to summarize all types of data, it is basically used to identify and determine nonconformities. All in all, the diagram could be quite useful to improve quality of medical device designs and additional information on the method is available in refs. [18, 20].

4.7 PROBLEMS

1. Compare medical device safety with reliability.
2. Discuss the following two types of medical device safety:
   - Conditional safety
   - Unconditional safety
3. Discuss the following two categories of medical device safety-related requirements imposed by governments and other regulatory bodies:
   - Safe design
   - Safe function
4. Discuss safety in the following three phases of a medical device:
   - Concept phase
   - Development phase
   - Deployment phase
5. List at least seven methods that can be used to perform safety analysis of medical devices.
6. What is the GMP regulation developed by the FDA?

7. Discuss at least four methods that can be used to improve quality of medical devices.
8. What are the main areas of the design quality assurance program recommended by the FDA?
9. Write a short essay on developments in the area of medical device safety.
10. Compare conditional safety with descriptive safety with regard to medical devices.

REFERENCES

1. Goetsch, D.L. *Occupational Safety and Health.* Englewood Cliffs, NJ: Prentice Hall, 1996.
2. LaDou, J., ed. *Introduction to Occupational Health and Safety.* Chicago: National Safety Council, 1986.
3. Dhillon, B.S. *Medical Device Reliability and Associated Areas.* Boca Raton, FL: CRC Press, 2000.
4. Dague, D.C. Quality: Historical Perspective. In *Quality Control in Manufacturing.* Warrendale, PA: Society of Automotive Engineers, 1981.
5. Fagan, M.D., ed. *A History of Engineering and Science in the Bell System, the Early Years (1875-1925).* New York: Bell Telephone Laboratories, 1974.
6. Fries, R.C. *Reliable Design of Medical Devices.* New York: Marcel Dekker, 1997.
7. Leitgeb, N. *Safety in Electro-medical Technology.* Buffalo Grove, IL: Interpharm Press, 1996.
8. Roland, H.E., Moriarty, B. *System Safety Engineering and Management.* New York: John Wiley & Sons, 1983.
9. Dhillon, B.S. *Engineering Safety: Fundamentals, Techniques, and Applications.* River Edge, NJ: World Scientific Publishing, 2003.
10. Goetsch, D.L. *Occupational Safety and Health.* Englewood Cliffs, NJ: Prentice Hall, 1996.
11. System Safety Analytical Techniques. Safety Engineering Bulletin No. 3, May 1971. Washington, DC: Electronic Industries Association,
12. Gloss, D.S., Wardle, M.G. *Introduction to Safety Engineering.* New York: John Wiley & Sons, 1984.
13. Hammer, W. *Product Safety Management and Engineering.* Englewood Cliffs, NJ: Prentice Hall, 1980.
14. Dhillon, B.S. *Advanced Design Concepts for Engineers.* Lancaster, PA: Technomic Publishing Company, 1998.
15. Dhillon, B.S. *Design Reliability: Fundamentals and Applications.* Boca Raton, FL: CRC Press, 1999.
16. Montanez, J. *Medical Device Quality Assurance Manual.* Buffalo Grove, IL: Interpharm Press, 1996.
17. Hooten, W.F. A Brief History of FDA Good Manufacturing Practices. *Medical Device & Diagnostic Industry* 18 (1996): 96-97.
18. Mears, P. *Quality Improvement Tools and Techniques.* New York: McGraw-Hill, 1995.
19. Besterfield, D.H., Besterfield-Michna, C., Besterfield, G.H., Besterfield-Sacre, M. *Total Quality Management.* Englewood Cliffs, NJ: Prentice Hall, 1995.
20. Sahni, A. Seven Basic Tools That Can Improve Quality. *Medical Device & Diagnostic Industry* April (1998): 89-98.
21. Bracco, D. How to Implement a Statistical Process Control Program. *Medical Device & Diagnostic Industry* March (1998) 129-139.

5 Medical Device Software Quality Assurance and Risk Assessment

5.1 INTRODUCTION

In the overall success of computer-controlled products, both hardware and software must work normally. In fact, software is an important element of modern medical devices. In 1984, approximately 80% of all major medical systems contained one or more computerized components and it was predicted that by 1990 almost all electrical medical devices would be composed of at least one computerized part.[1-3] Furthermore, over the years, the size of software used in medical devices has increased quite dramatically; for example, the software of a typical cardiac rhythm management device is made up of approximately half a million lines of code.[4]

During the period 1983 to 1985, according to the U.S. Food and Drug Administration (FDA), a total of 41 medical product recalls were concerned with software.[5] Needless to say, the quality assurance of medical device software is as important as the quality assurance of medical device hardware.

The history of modern risk analysis may be traced back to 1792, when mathematician Pierre-Simon de Laplace estimated the probability of death with and without smallpox vaccination.[6] However, in the twentieth century, two of the important factors responsible for the conceptual developments of risk analysis in the United States were the concerns about the safety of nuclear power plants and the establishment of various government agencies/bodies; for example, the Environmental Protection Agency and Occupational Safety and Health Administration. Risk may be defined as a measure of the probability and security of a negative effect to equipment, property, environment, or health.

In the area of medical devices/equipment, risk assessment and control is essential from various different perspectives, including economics and safety. This chapter presents various important aspects of medical device software quality assurance and risk assessment.

5.2 MEDICAL DEVICE SOFTWARE FAILURE EXAMPLES

Over the years there have been many medical device failures due to various types of software-related problems. Some of these were as follows:

- Two persons died in heart pacemaker-related incidents due to software problems.[3]

- Two patients died and a third was badly injured because of software-related faults in a computer-controlled therapeutic radiation machine called Therac 25.[3,7,8]
- An infusion pump delivered at the maximum rate instead of intended rate subsequent to entering certain correct data, because of a software-related problem.[3]
- A multiple-patient monitoring system failed to store collected data with the right patient because of a software error.[2,3,9,10]
- The London Ambulance Service's newly installed computer-aided dispatch system failed because of software-related problems, resulting in the incorrect ambulance being sent to an incident.[11]
- A diagnostic laboratory instrument containing a software fault resulted in wrong reports of patient data.[2,3]

5.3 CLASSIFICATIONS OF MEDICAL SOFTWARE

In the United States, the FDA classifies medical software under two categories, as shown in Figure 5.1.[12] These are stand-alone software and software that is a component, part, or accessory to a device. The stand-alone software could be treated as a medical device; thus, it is subject to all applicable FDA medical device statutory and regulatory provisions. Four typical examples of the stand-alone medical software are shown in Figure 5.2.

The software that is a component, part, or accessory to a device is regulated according to the parent device requirements, unless it is separately classified. More specifically, software as a component/part may be stated as any software intended to be included as a component of the finished, packaged, and labeled medical device. Some examples of such a device are ventilators, infusion pumps, and x-ray systems.

Similarly, the software as an accessory may be stated as a unit intended to be attached or used in conjunction with another finished device, even though it could be promoted or sold totally as an independent unit. Four examples of such software are shown in Figure 5.3.[12]

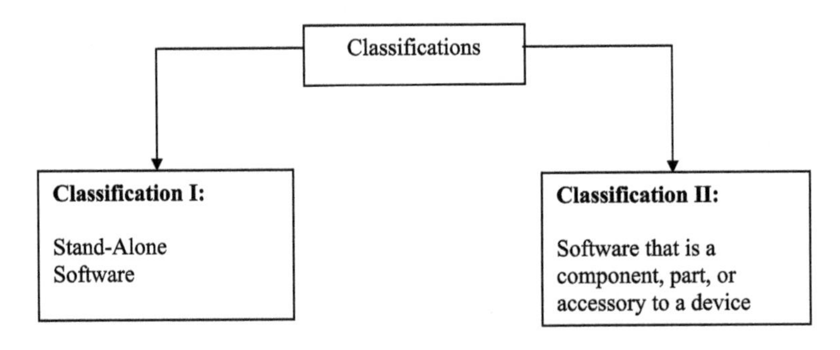

FIGURE 5.1 Medical software classification.

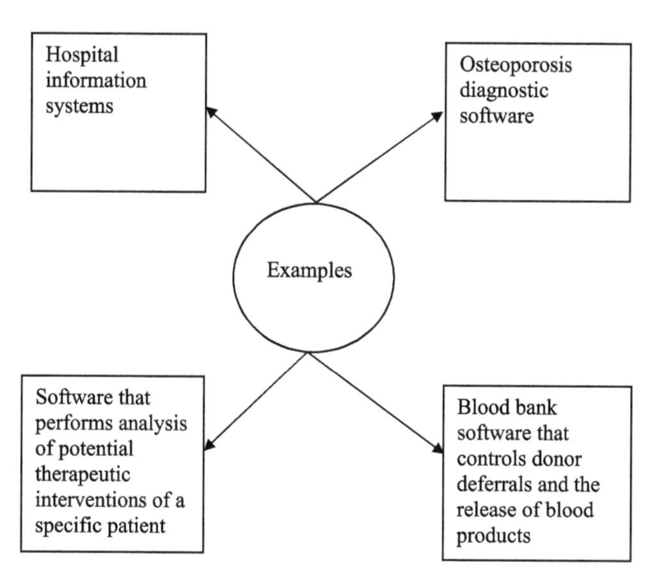

FIGURE 5.2 Examples of stand-alone software.

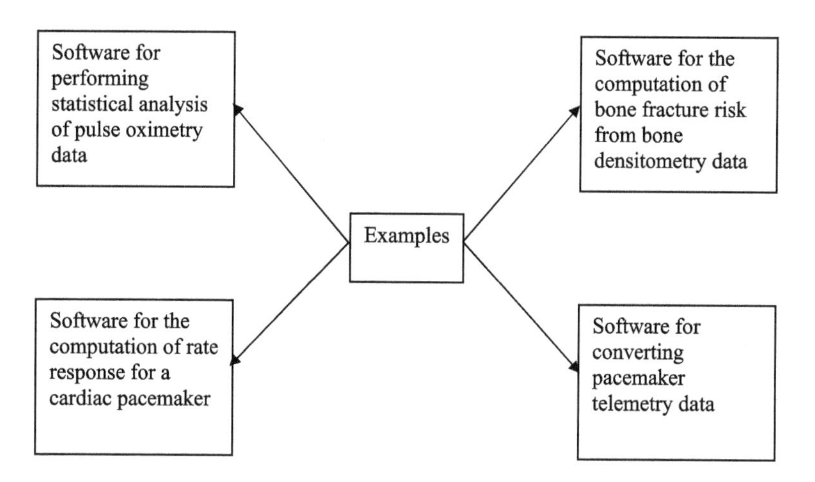

FIGURE 5.3 Examples of accessory software.

5.4 FRAMEWORK FOR DEFINING SOFTWARE QUALITY ASSURANCE PROGRAMS IN MEDICAL DEVICE MANUFACTURING ORGANIZATIONS

As the use of software is growing rapidly in medical device manufacturing organizations, many of these organizations implement software quality assurance programs to assure the conformance of software products and development processes to specified requirements. The following framework of eight steps is considered to be quite useful to establish software quality assurance programs in such organizations[13]:

Step 1: Collect requirements. This step is concerned with gathering information related to the program. The information includes items such as internal and external requirements, organizational culture, and industry standards.

Step 2: Establish a plan. This step is concerned with developing a plan that includes items such as major activities to be accomplished, expected completion dates, and required resources.

Step 3: Develop mission statement of software quality assurance. This step is concerned with establishing a mission statement that is traceable to the organization's mission statement and its approval by the upper management.

Step 4: Develop a policy and standard. This step is concerned with establishing policy and standard. The policy focuses on various important organizational concerns, including software quality assurance having the authority to take corrective actions to carry out improvements in the quality of the software development processes and products and the authority to perform independent evaluations of software development processes and products. The standard provides appropriate description supporting the policy.

Step 5: Highlight all relevant activities. This step is concerned with the identification of specific activities including the analysis of the current activities (if any) being performed and the identification of the most valuable ones, as well as the identification of additional activities appropriate to meet the mission statement.

Step 6: Develop appropriate operating procedures. This step is concerned with developing standard operating procedures for providing adequate information to perform software quality assurance activities effectively.

Step 7: Train and promote the program. This step is concerned with providing necessary training and promoting the program.

Step 8: Review the program. This step is concerned with evaluating the program for the purpose of making appropriate improvements.

5.5 SOFTWARE DESIGN, CODING, AND TESTING

Design, coding, and testing are those elements of the software life cycle that greatly influence software's effectiveness in real-life use. Thus, they are critical to produce safe and reliable medical device software. These three items are discussed in detail below, separately.[14-16]

5.5.1 Software Design

Various studies conducted over the years indicate that approximately 45% of software system errors tend to be design errors.[14,17] Software design may be divided into two groups, as follows[18]:

- **Top-level design.** It has many important elements including methodology selection, language selection, software architecture, software review, risk analysis, design alternatives and trade-offs, software requirements traceability matrix, and object-oriented design.
- **Detailed design.** It has major components such as design methods, design support tools, performance predictability and design simulation, and module specification.

Software design representation approaches and design methods are described below.

Design Representation Approaches

In software design, various design representation schemes are used. Some of these are flowchart, Warnier-Orr diagram, data structure chart, Chapin charts, and hierarchy plus input-process-output (HIPO) diagram. Additional information on these schemes is available in refs. [14, 19-21].

Software Design Methods

Three commonly used software design methods are modular design, structured programming, and top-down programming. Each of these methods is described below, separately.

Modular Design

Modular design is concerned with decomposing complex medical software development jobs into various modules. A module may be described as a self-contained, modest-sized subprogram performing independently on one specific function and if it is totally removed from the system then it only disables its unique function—nothing more. Nonetheless, two major criteria associated with the modular design are functional independence restriction and module size restriction.

The functional independence restriction is concerned with defining the logic scope a module may cover, whereas the module size restriction is concerned with defining maximum and minimum number of program language statements in a given module.

Some of the main advantages of the modular design approach are as follows[14]:

- Easy to write, debug, and test software
- Cheaper to make changes and corrections during the deployment phase
- More reliable software

In contrast, two of the main disadvantages of the method are as follows[14]:

- Requires more effort during design
- May require more memory space

Structured Programming

Structured programming may be described as coding that avoids the use of GoTo statements and program design that is modular and top-down.[22] For the average programmer, some easy to use rules associated with structured programming are: use a language and compiler that have structured programming capabilities for implementation, ensure that each module has only one entry and one exit, ensure that all statements with the inclusive of subroutine calls are commented, and restrict the use of GoTo statements in modules.[23]

Some of the advantages of structured programming are as follows:

- An effective tool for increasing programmer productivity
- Maximization of the amount of code that can be reused in redesign effort
- Useful to localize an error
- Useful to understand the program design by the designers and others

The main disadvantages of structured programming include: it requires additional effort because many programming languages lack certain control concepts necessary for implementing the structured programming objective, and it may require additional memory and running time.

Top-Down Programming

This software design method may be described as a decomposition process that directs attention to the program flow of control or to the program control structure. Top-down programming begins with a module representing the total program and then the module is decomposed into a number of subroutines. These subroutines are broken down further, until their broken down elements are easily comprehensible and simple to work with.[24]

Some of the advantages of top-down programming are lower testing cost, better quality software, and a more readable form of the end product.

5.5.2 Software Coding

The basic goal of the software development implementation phase is to translate design specifications into an effective source code. Thus, the source code and internal documentation are written by paying close attention to the fact that it is quite straightforward to verify conformance of the code to its specification. In turn, this will make it easy to debug, test, and modify software. Factors such as structured coding methods, appropriate supporting documents, good internal comments, and good coding style are quite useful to enhance clarity of the source code.

In particular, some of the useful guidelines for achieving good coding style or practices are: review each line of code, emphasize that code listing are public assets, require coding sign-offs, reward good code practices.[18,25,26]

5.5.3 Software Testing

Software testing accounts for a substantial proportion of the software development process and it may be described as the process of executing a program with the intention of finding errors or bugs. There are many factors behind the testing of medical device software, including market competition, regulatory pressure from bodies such as the FDA, and increased concern for safety as a consequence of poorly tested software that could be very devastating for both producers and patients.[27,28]

The development of a good test plan is essential in order to have an effective testing program. The plan must consider factors such as duration of the test phase, error classification scheme, function test needs, test strategy, testing criteria to be satisfied, and test result documentation.[29] The actual software testing function may be divided into two categories: manual testing and automated testing.[28] Both categories are described below, separately.

Manual Software Testing

There are many approaches to manual software testing. Some of the traditional ones are shown in Figure 5.4.[28]

White-box testing is used when it is essential to verify the internal functioning of the device/equipment software. It allows the tester to look inside the device/equipment and develop appropriate tests for finding weak spots in the internal logic of the program.

Error testing is used to ensure that the device/equipment performs properly in abnormal conditions, such as power surges/outages, internal part failures, or incomprehensible values generated by peripheral devices.

Free-form testing is used in situations when no formal test plan can discover all important software bugs. Thus, free-form testing may be described as a process in which testers with a varying degree of knowledge use the device/equipment in unexpected and unconventional ways in order to provoke failures.

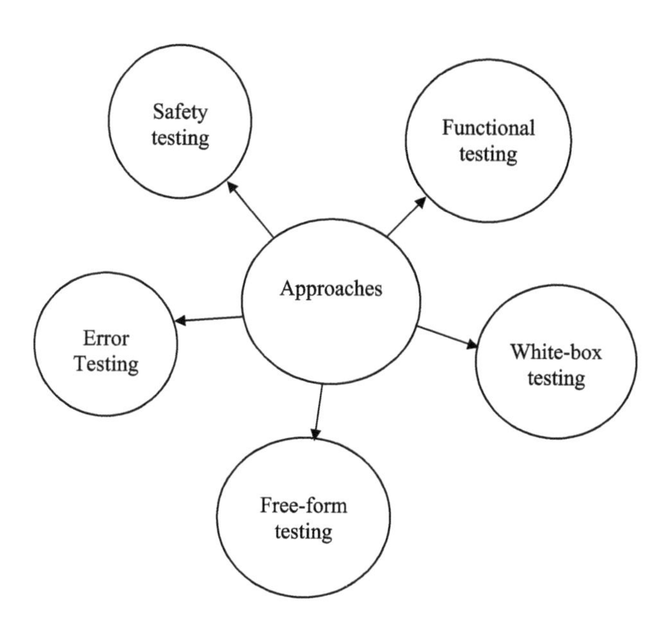

FIGURE 5.4 Approaches to manual software testing.

Safety testing is used to focus test efforts on conditions potentially dangerous to the user or patient. Functional testing is used to direct test efforts on what the device/equipment is expected to do and it may simply be called an orderly process. This type of testing usually receives most attention and the greatest number of test cases because each and every function is tested exhaustively.

Automated Software Testing

Automated software testing uses computer technology for testing software so that the test process is more accurate and complete, faster, and is backed up with better quality documentation. More specifically, automated software test systems use computer technology for various purposes, including to stimulate the test target, to control the complete process, and to record results.

In the case "to stimulate the test target," automated testing augments through computer simulations: human, environmental, and internal stimulation of the target. Similarly, in the case "to control the complete process," automated testing totally replaces manual test-related plans with computer-based test programs. Finally, in the case "to record results," automated testing replaces handwritten test notes with computer-based records.

5.6 SOFTWARE METRICS

Software metrics are used to measure software complexity, and over the years various types of software metrics have been developed. These metrics are used in areas such as goal setting, project planning and managing, enhancing customer confidence, and improving quality and productivity.[18,30]

In selecting metrics for use, careful consideration must be given to factors such as support (they) the determination of the software in line with the given requirements, usefulness to the specific objectives of the program, and derivation (their) from the program requirements. Two widely used software metrics are as follows:

- McCabe's complexity (metric I)
- Halstead measures (metric II)

Both of these metrics are described below, separately.

5.6.1 Metric I: McCabe's Complexity

As this metric is based on the complexity of the flow-of-control in the system, the metric relates the complexity to the software/program measure of structure. Needless to say, McCabe's complexity is derived from classical graph theory and is expressed as follows[6,30-32]:

$$CN = \mu - \theta + 2y \tag{5.1}$$

where

CN is the complexity number.
μ is the number of edges in the program.
θ is the total number of nodes or vertices.
y is the total number of separate tasks or connected components.

Example 5.1

Assume that in Equation (5.1) we have the following specified data values:

$$\theta = 8$$
$$y = 3$$
$$\mu = 3$$

Calculate the complexity number and comment on the result.

By inserting the specified data values into Equation (5.1), we obtain:

$$CN = 3 - 8 + 2\,(3)$$
$$= 1$$

The value of the complexity number is equal to unity or 1. It simply means that one can expect high software reliability. In general, it may be added that the higher the value of the complexity number, the more difficult it will be to develop, test, and maintain software under consideration; thus the lower its reliability.

5.6.2 Metric II: Halstead Measures

These measures basically provide a quality measure of the software development process by considering the following two factors[18,30]:

- Number of distinct operators (i.e., types of instructions)
- Number of distinct operands (i.e., constraints and variables)

The measures are named after Professor Maurice Halstead of Purdue University and include software vocabulary, program length, software volume, and the potential volume.[33,34] All these measures are presented below, separately.

Software Vocabulary

This is expressed by:

$$SV = \lambda + \alpha \tag{5.2}$$

where

SV is the vocabulary of the software.
α is the total number of distinct operands.
λ is the total number of distinct operators.

Program Length

This is defined by:

$$PL = m + n \tag{5.3}$$

where

PL is the program length.
m is the total occurrences of operands.
n is the total occurrences of operators.

Software Volume

This is expressed by:

$$VS = PL \log_2 SV \tag{5.4}$$

where

VS is the volume of the software.

Potential Volume

This is defined by

$$VS^* = (2 + \alpha)\left[\log_2 (2 + \alpha)\right] \tag{5.5}$$

where

VS^* is the potential volume.

5.7 RISK MANAGEMENT DEFINITION AND PROGRAM STEPS

Although there are various definitions of risk management in the published literature, it may be defined as the systematic application of management policy, procedures, and practices to identify, analyze, control, and monitor risk.[35]

The following steps are associated with a risk management program[6,36]:

Step 1: Establish requirement definition and mechanisms for achieving it.
Step 2: Outline related responsibilities and accountability.
Step 3: Clearly identify what requires authorization and responsibility for handling it.
Step 4: Define the skills and knowledge necessary for system implementation and a provision for providing training to those without the required skills.
Step 5: Develop appropriate documentation for demonstrating conformance to specified policies and procedures.
Step 6: Develop and incorporate essential measures for cross-checking and verifying that all procedures are being followed properly.
Step 7: Conduct appropriate verification that the systems are properly in place and operating normally.

5.8 FACTORS IN MEDICAL DEVICE RISK ASSESSMENT

Factors that are critical to medical device risk assessment relate both to the device as well as its usage. The critical factors that relate to the device and its usage are shown in Figures 5.5 and 5.6, respectively.[37] Some of these factors are described below, separately.

5.8.1 Design and Manufacturing

A medical device could be hazardous if improper attention is paid to design aspects influencing device performance or it is poorly manufactured. Even good manufacturing practices do not eradicate inherently bad designs; in fact, they may simply help to produce a "bad design" faithfully. On the other hand, inadequate attention to a manufacturing quality assurance program may result in a defective device.

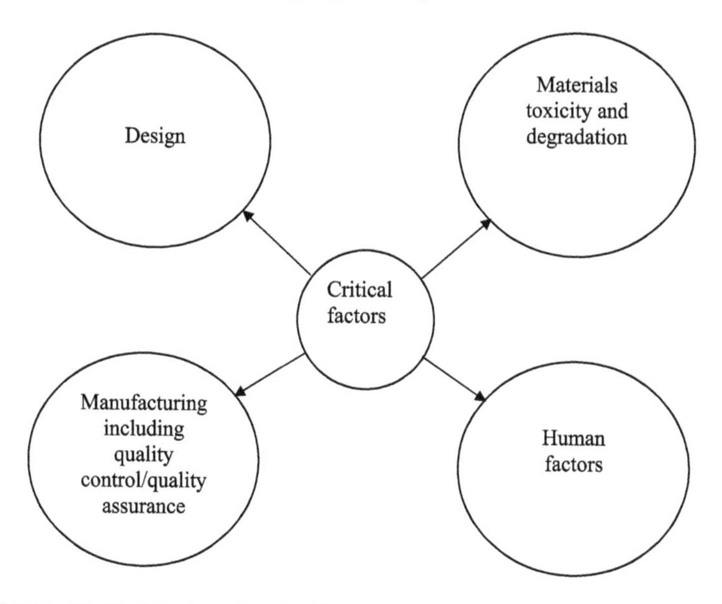

FIGURE 5.5 Medical device-related critical factors.

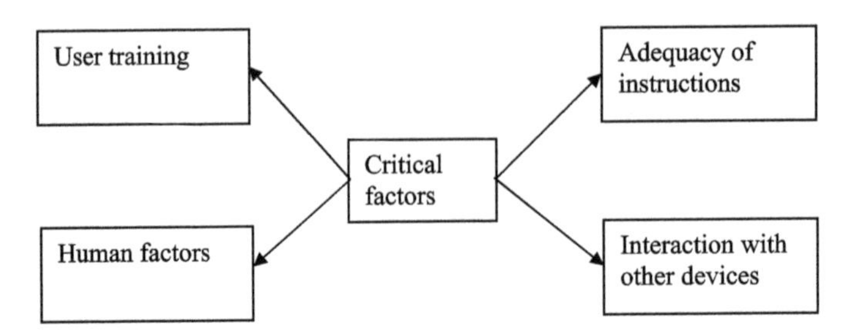

FIGURE 5.6 Medical device usage-related critical factors.

In both the cases, i.e., faulty design and flawed design implementation, the hazard type may not be that different, but it is quite essential that risk assessment make a clear-cut distinction between the two, or determine if both are actually involved at a reasonable level. A good knowledge of this type of information is absolutely necessary because the approaches to risk management in each case will be quite different.

5.8.2 Materials Toxicity and Degradation

The selection of proper material is an important element in a good design, and toxicity is an important factor to determine the risk of a material used in a medical device developed for implant. Past experiences indicate that it could be a challenging task to determine the material used in a device because a device can contain contaminant material, catalytic material used in synthesis, the material making up the primary molecular structure, etc.

Nonetheless, the tissue environment could be an important influencing factor for leachability that can affect whether the toxic agent remains locally or is distributed throughout the system. Furthermore, the tissue environment could be quite hostile as well, as it can initiate, promote, or accelerate the material degradation.

Usually, the hostility of a specific environment depends on the following two factors:

- The type of material
- The length of time the material remains in the environment

Thus, risk assessment must consider factors such as these as well as a good judgment.

5.8.3 Human Factors

These are those medical device use conditions and elements that influence interactions between the device and the user. Past experiences indicate that poor attention given to human factors in the early generation of medical devices has resulted in various types of problems. All in all, it is added that without seriously considering human factors as an important element of risk assessment, it would be very difficult to consider the effective risk management approach for medical devices/equipment.

5.8.4 Interaction With Other Devices

Past experiences indicate that from time to time medical devices are used within close proximity to each other. Although such scenarios may have been seriously considered in the device design, the worst case design allowance could sometimes be bypassed by the ingenuity of the device users. All in all, risk assessment must take into consideration factors such as electromagnetic interference from one device to another and the possibility of users interchanging incompatible device parts during the repair process.

5.9 INTEGRATING RISK ASSESSMENT INTO MEDICAL DEVICE DESIGN CONTROL

Over the years, errors or deficiencies in device design have been an important factor in voluntary device recalls in the United States.[35] Nowadays, the FDA expects that risk analysis must be fully integrated with the device design control effort.[38]

The design control activities, in which risk assessment can be integrated, are discussed with respect to risk assessment below.[35]

- **Project planning.** In this activity, the risk management plan should be made an element of the project management/product development plan. The plan should incorporate a strategic mechanism to identify and control risk in the product development life cycle.
- **Design input.** In this activity, the pertinent design-related inputs are the existing safety requirements and standards identified in risk assessments.
- **Design out.** In this activity, the essential design outputs are the risk reduction measures introduced into device design.
- **Design transfer.** In this activity, each formal design review should address risk assessment activities, as necessary, to assure that specified actions are assigned/monitored effectively prior to the design transfer.
- **Design verification.** In this activity, one affirms that relevant safety requirements are considered by the risk reduction measures in the device design.
- **Design validation.** In this activity, all efforts should be directed to demonstrate that relevant safety requirements can be consistently and effectively satisfied by considering all patient/user needs and intended applications.

5.10 MEDICAL DEVICE RISK ASSESSMENT-RELATED DATA

Over the years, professionals working in the area of health care risk assessment have come up with estimates concerning various causes and events. A sample of these estimates, directly or indirectly related to medical devices, is presented in Table 5.1.[39] Data on incidence of adverse events and negligence in hospitalized patients and on other areas, directly or indirectly related to medical device risk assessment, are available in refs. [6, 40-41].

Table 5.1 Average Loss of Life Expectancy Due to Medical-Related Causes and All Catastrophes Combined

Cause	Average Life Expectancy Loss in Days
Legal drug misuse	90
Illicit drugs	18
Medical x-rays	6
Electrocution	5
All catastrophes combined	35

5.11 PROBLEMS

1. List at lest five examples of medical device software failure.
2. Discuss two main classifications of medical device software.
3. Discuss the following:
   - Structured programming
   - Modular design
   - Top-down programming
4. Discuss the three most useful approaches to manual software testing.
5. What are the Halstead measures?
6. Define risk management.
7. Discuss the steps associated with a risk management program.
8. Discuss design control activities with respect to risk assessment.
9. What is McCabe's complexity?
10. Discuss factors that are critical to medical device risk assessment.

REFERENCES

1. Anbar, M., ed. *Computers in Medicine.* Rockville, MD: Computer Science Press, 1987.
2. Bassen, H., Silberberg, J., Houston, F., Knight, W., Christman, C., Greberman, M. Computerized Medical Devices: Usage Trends, Problems, and Safety Technology. Proceedings of the 7th annual conference of the IEEE, Engineering in Medicine and Biology Society, on Frontiers of Engineering and Computing in Health Care, 1985, 180–185.
3. Schneider, P., Hines, M.L.A. Classification of Medical Software. Proceedings of the IEEE Symposium on Applied Computing, 1990, 20–27.
4. Vishnuvajjala, R.V, Subramaniam, S., Tsai, W.T., Elliott, L., Mojdehbaksh, R. Run-Time Assertion Scheme for Safety Critical Systems. Proceedings of the 9th IEEE Symposium on Computer-Based Medical Systems, 1996, 18–23.
5. Estrin, N.F., ed. *The Medical Device Industry: Science, Technology, and Regulation in a Competitive Environment.* New York: Marcel Dekker, 1990.
6. Dhillon, B.S. Medical Device Reliability and Associated Areas. Boca Raton, FL: CRC Press, 2000.
7. Gowen, L.D., Yap, M.Y. Traditional Software Development's Effects on Safety, Proceedings of the 6th annual IEEE Symposium on Computer-Based Medical Systems, 1993, 58–63.
8. Joyce, E. Software Bugs: A Matter of Life and Liability. *Datamation* 33 (1987): 88–92.

9. Neumann, P.G. Risks to Public. *ACM SIGSOFT Software Engineering Notes* 11 (1986): 34–35.
10. Neumann, P.G. Some Computer-Related Disasters and Other Egregious Horrors. Proceedings of the 7th annual conference of the IEEE Engineering in Medicine and Biology Society, 1985, 1238–1239.
11. Shaw, R. Safety-Critical Software and Current Standards Initiatives. *Computer Methods and Programs in Biomedicine* 44 (1994): 5–22.
12. Onel, S. Draft Revision of FDA's Medical Device Software Policy Raises Warning Flags. *Medical Device & Diagnostics Industry* 19 (1997): 82–91.
13. Linberg, K.R. Defining the Role of Software Quality Assurance in a Medical Device Company. Proceedings of the 6th annual IEEE Symposium on Computer-Based Medical Systems, 1993, 278–283.
14. Dhillon, B.S. *Reliability in Computer System Design.* Norwood, NJ: Ablex Publishing Corporation, 1987.
15. Fisher, K.F. A Methodology for Developing Quality Software. Proceedings of the annual American Society for Quality Control Conference, 1979, 364–371.
16. Dunn, R., Ullman, R. *Quality Assurance for Computer Software.* New York: McGraw-Hill, 1982.
17. Joshi, R.D. Software Development for Reliable Software Systems. *Journal of Systems and Software* 3 (1983): 107–121.
18. Fries, R.C. *Reliable Design of Medical Devices.* New York: Marcel Dekker, 1997.
19. Fairly, R.E. Modern Software Design Technique. Proceedings of the Symposium on Computer Software Engineering, 1976, 111–131.
20. Peters, L.J., Tripp, L.L. Software Design Representation Schemes. Proceedings of the Symposium on Computer Software Engineering, 1976, 31–56.
21. Shooman, M.L. *Software Engineering: Design, Reliability, and Management.* New York: McGraw-Hill, 1983.
22. Rustin, R. *Debugging Techniques in Large Systems.* Englewood Cliffs, NJ: Prentice Hall, 1971.
23. Wang, R.S. Program with Measurable Structure. Proceedings of the American Society for Quality Control Conference, 1980, 389–396.
24. Koestler, A. *The Ghost in the Machine.* New York: Macmillan, 1967.
25. McConnell, S.C. *Code Complete: A Practical Handbook of Software Construction.* Redmond, WA: Microsoft Press, 1993.
26. Maguire, S.A. *Writing Solid Code: Microsoft's Techniques for Developing But-Free C Programs.* Redmond, WA: Microsoft Press, 1993.
27. Jorgens, J. *The Purpose of Software Quality Assurance: A Means to an End, in Developing Safe, Effective, and Reliable Medical Software.* Arlington, VA: Association for the Advancement of Medical Instrumentation, 1991, 1–6.
28. Weide, P. Improving Medical Device Safety with Automated Software Testing. *Medical Device & Diagnostics Industry* 16 (1994): 66–79.
29. Kopetz, H. *Software Reliability.* London: Macmillan Press Ltd., 1979.
30. Fries, R.C. *Reliability Assurance for Medical Devices, Equipment, and Software.* Buffalo Grove, IL: Interpharm Press, 1991.
31. McCabe, T. A Complexity Measure. *IEEE Transactions on Software Engineering* 2 (1976): 308–320.
32. Shooman, M.L. *Software Engineering.* New York: McGraw-Hill, 1983.
33. Halstead, M.H. Software Physics: Basic Principles. Report RJ1582. York Town, NY: IBM Research Laboratory, May 20, 1975.
34. Halstead, M.H. *Elements of Software Science.* New York: Elsevier, 1977.
35. Knepnell, P.L. Integrating Risk Management with Design Control. *Medical Device & Diagnostics Industry* October (1998) 83–91.

36. Ozog, H. Risk Management in Medical Device Design. *Medical Device & Diagnostics Industry* October (1997): 112–120.
37. Anderson, F.A. Medical Device Risk Assessment. In *The Medical Device Industry: Science, Technology, and Regulation in a Competitive Environment*, edited by N.F. Estrin, 487–493. New York: Marcel Dekker, 1990.
38. Design Control Guidance for Medical Device Manufacturers. Rockville, MD: U.S. Food and Drug Administration, March 1997.
39. Cohen, B.L., Lee, I.S. A Catalog of Risks. *Health Physics* 36 (1979): 707–708.
40. Brennan, T.A., Leape, L.L., Laird, N.M., et al. Incidence of Adverse Events and Negligence in Hospitalized Patients. *New England Journal of Medicine* 324 (1991): 370–376.
41. McCormick, N.J. *Reliability and Risk Analysis*. New York: Academic Press, 1981.

6 Medical Device Maintenance and Sources for Obtaining Medical Device-Related Failure Data

6.1 INTRODUCTION

Maintenance is an important element of the engineering equipment life cycle. In fact, according to various studies, normally much more money is spent to maintain equipment over its entire lifespan than its original acquisition cost. For example, according to a United States Air Force (USAF) study, equipment repair and maintenance costs are approximately 10 times greater than its original procurement cost.[1] Needless to say, the maintenance of medical equipment or devices is as important as their design and development.

As in the case of general engineering equipment, failure data are the backbone of medical device-related reliability studies. They are the final proof of the degree of the reliability-related efforts expended during design and manufacture of medical devices/equipment. Thus, failure data provide invaluable information to professionals involved with the design, manufacture, and maintenance of future medical devices/equipment.

This chapter presents various important aspects of medical device maintenance and sources for obtaining medical device-related failure data.

6.2 MEDICAL EQUIPMENT CLASSIFICATIONS

A wide range of equipment is used in the health care system. It may be classified under the following six categories[2,3]:

- **Imaging and radiation therapy.** This category includes radiation therapy equipment and the devices used to image patient anatomy. Three examples of such equipment are linear accelerators, x-ray machines, and ultrasound devices.
- **Life support and therapeutic.** This category includes equipment used to apply energy to the patient. Some examples of such equipment are powered surgical instruments, lasers, ventilators, and anesthesia machines.

- **Patient diagnostic.** This category includes devices connected to patients and used for collecting, recording, and analyzing information concerning patients. Two examples of such devices are physiologic monitors and spirometers.
- **Patient environmental, transport.** This category includes patient beds and equipment used to transport patients or improve patient environment. Some examples of such equipment are wheelchairs, patient room furniture, and examination lights.
- **Laboratory apparatus.** This category includes devices employed in the preparation, storage, and analysis of in vitro patient specimens. Laboratory refrigeration equipment, centrifuges, and laboratory analyzers are typical examples of such devices.
- **Miscellaneous.** This category includes all those medical equipment/devices that are not included in the other five categories.

6.3 MEDICAL EQUIPMENT MAINTENANCE INDEXES

Over the years various types of indexes have been used to measure different aspects of the maintenance function. This section presents three broad indexes and three indexes specifically used in the area of health care.

6.3.1 Broad Indexes

A number of broad maintenance indexes are used in the industrial sector to monitor the effectiveness of the maintenance activity in an organization. Three of these indexes considered useful for application in health care organizations are presented below.[4-8]

Index I

This index relates the total maintenance cost to the total investment in plant and equipment and is expressed by:

$$IX_1 = \frac{C_m}{TI} \tag{6.1}$$

where

IX_1 is the index parameter.
C_m is the total maintenance cost.
TI is the total investment in plant and equipment.

The average values of IX_1 for steel and chemical industries are approximately 8.6% and 3.8%, respectively.

Index II

This is expressed by:

$$IX_2 = \frac{C_m}{TSR} \quad (6.2)$$

where

IX_2 is the index parameter.
TSR is the total sale revenue.

Generally, average expenditure for the maintenance activity for all industry is around 5% of the sale revenue.[8] However, there is a wide variation within the industrial sector. For example, the average values of the index, IX_2, for chemical and steel industries are 6.8% and 12.8%, respectively.

Index III

This index relates the total maintenance cost to the total output by the organization in question and is expressed by:

$$IX_3 = \frac{C_m}{TOT} \quad (6.3)$$

where

IX_3 is the index parameter.
TOT is the total output expressed in tons, megawatts, gallons, etc.

6.3.2 Medical Equipment-Specific Indexes

Over the years the Association for the Advancement of Medical Instrumentation (AAMI) has developed a number of common, standardized cost and quality indexes to help medical technology managers compare repair and maintenance services among organizations involved with medical equipment,[2] consequently, reducing repair and maintenance costs and enhancing the effectiveness of repair/maintenance services. Three such indexes are presented below.

Index IV

This index basically provides repair requests accomplished per medical device/equipment; thus, it is analogous to device/equipment repair rate. In other words, the index measures how frequently the customer has to request service per medical device/equipment. The index is defined by:

$$IX_4 = \frac{TNRR}{\alpha} \quad (6.4)$$

where

IX_4 is the index parameter or the number of repair requests accomplished per medical device/equipment.
α is the total number of medical devices/equipment.
$TNRR$ is the total number of repair/requests.

According to one study, the value of IX_4 varied from 0.3 to 2.0 with a mean of 0.8.[2]

Index V

This is defined by:

$$IX_5 = \frac{TAT_a}{\theta} \tag{6.5}$$

where

IX_5 is the index parameter or the average turnaround time per repair.
θ is the number of work orders or repairs.
TAT_a is the total turnaround time.

This index measures how much time elapses from a customer request, until the malfunctioning medical device/equipment is repaired and put back into full service.

According to a study reported in ref. [2], turnaround time per repair in five hospitals varied from 35.4 hours to 135 hours. The average value of this time for these five hospitals was 79.5 hours.

Index VI

This is defined by:

$$IX_6 = \frac{C_S}{C_p} \tag{6.6}$$

where

IX_6 is the index parameter.
C_p is the procurement cost or the cost at the time of purchase of equipment.
C_S is the service cost or the total of all parts, materials, and labor costs for scheduled and unscheduled service. This cost also includes in-house, vendor, prepaid contracts, and maintenance insurance expenses.

According to a study reported in ref. [2], the value of IX_6 varied from 2.1% to 5.5% with a mean of 3.9%. In addition, the average values of IX_6 for six classifications of medical equipment (laboratory apparatus, imaging and radiation therapy, patient diagnostic, life support and therapeutic, patient environmental and transport, and miscellaneous) were 5.1%, 5.6%, 2.6%, 3.5%, 4.4%, and 2.6%, respectively.

Some of the advantages of this index are as follows[2,3]:

- Easy to compare across equipment types
- Takes into consideration all types of service costs
- Can be used with incomplete data

Similarly, some disadvantages of the index are as follows[2,3]:

- Allows no compensation for the age of the equipment
- Requires a standard definition for the pricing of in-house service
- Possesses no mechanism for adjusting wage rates by geographic region

6.4 MEDICAL EQUIPMENT COMPUTERIZED MAINTENANCE MANAGEMENT SYSTEMS

Nowadays, usually hospital clinical engineering departments use various types of computerized maintenance management systems (CMMS) for collecting, storing, analyzing, and reporting data on the repair and maintenance performed on medical devices/equipment. In turn, these data are used in areas such as equipment management, work order control, quality improvement activities, reliability and maintainability studies, and cost control.

Today, there are many commercially available CMMS that can be used by clinical engineering departments.[9] In order to procure an effective CMMS, a careful consideration is necessary during the selection process. In the past, the use of a similar process to the prepurchase evaluation of a medical device/equipment has proven to be quite useful in selecting a commercial CMMS. The major steps of this process are shown in Figure 6.1.[9]

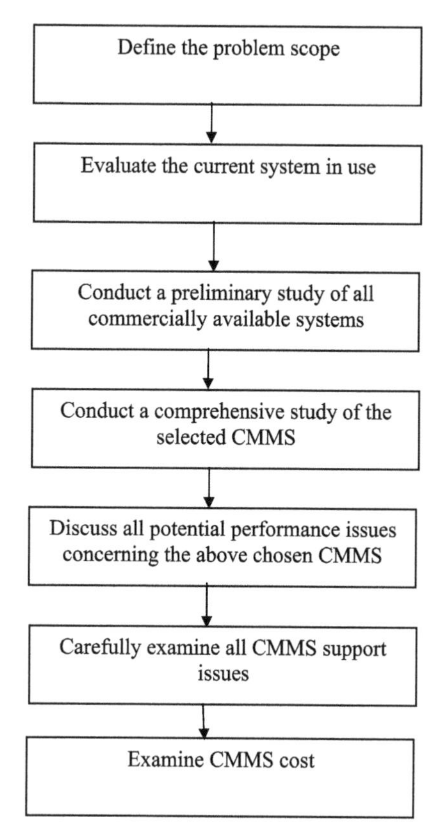

FIGURE 6.1 Major steps involved in selecting commercially available computerized maintenance management systems (CMMS) for use by hospital clinical engineering departments.

The first step "Define the problem scope" is basically concerned with defining the problem scope for the

Table 6.1 Questions for Evaluating the Effectiveness of the Current Computerized Maintenance Management Systems (CMMS) Being Used

Questions
1. How is the current data being used?
2. What are the types of data currently available?
3. What are the projected requirements?
4. How does the current data reach the clinical engineering department?
5. What are the legitimate sources of the current data?
6. What are the elements or components the present system does not have?
7. Is there any need to specify other special or unique requirements to satisfy potential needs effectively?
8. Are any special or unique requirements currently in progress?

clinical engineering department. Thus, an example of the problem scope could be "the clinical engineering department needs to acquire a CMMS, so that medical device/equipment repair data are collected, stored, analyzed, and reported more efficiently." The second step "Evaluate the current system in use" is basically concerned with evaluating the effectiveness of the existing system and it could be accomplished by answering questions such as presented in Table 6.1.

The third step "Conduct preliminary study of all commercially available systems" is concerned with evaluating the commercially available CMMS, simply by going through the marketing literature and understanding the general philosophy and focus of each system. The final result of this study can be utilized in developing a preliminary budget and determining which system to focus on for comprehensive examination.

The fourth step "Conduct a comprehensive study of the selected CMMS" is basically concerned with evaluating the chosen one or more CMMS, with respect to identified requirements. In the evaluation process, careful consideration is given to the flexibility of CMMS. The fifth step "Discuss all potential performance issues concerning the above chosen CMMS" is basically concerned with having discussions on potential performance issues with organizations of similar size and scope, who have used the CMMS under consideration. The results of the discussions with these organizations are normally used in determining whether the performance of systems under consideration will adequately meet the in-house needs.

The sixth step "Carefully examine all CMMS support issues" is basically concerned with evaluating the vendor support services. These services may be classified under two groups: initial start-up support and continuing support. The initial start-up support includes items such as user training, initial system documentation, initial assistance for software and hardware installation, and initial data entry of codes and inventory data. Similarly, the continuing support includes items such as periodic software enhancements and "bug" fixes, electronic bulletin board, and telephone support (i.e., the 800 number).

The seventh or the last step "Examine CMMS cost" is basically concerned with developing preliminary budget and cost estimates. The cost estimates can be utilized to perform life cycle cost analysis of the selected "finalists." Some examples of the costs involved are initial CMMS software cost, cost of initial training, annual software support cost, cost of initial network operating system license, and computer hardware costs.

6.5 MATHEMATICAL MODELS FOR MEDICAL EQUIPMENT MAINTENANCE

Over the years many mathematical models have been developed for application in the area of engineering maintenance.[8] This section presents two such models considered useful for application in medical equipment maintenance.[10-13]

6.5.1 Model I

This model can be used to determine the optimum time between item/equipment replacements. The model assumes that the equipment/item average annual cost is composed of three main components: average maintenance cost, average operating cost, and average investment cost. Thus, the model minimizes the equipment/item average annual cost with respect to its life.

Mathematically, the equipment /item average annual cost, EAAC, is expressed by:

$$EAAC = EOC_f + EMC_f + \frac{EIC}{t} + \frac{(t-1)}{2}(i+j) \tag{6.7}$$

where

EOC_f is the equipment/item operating cost for the first year.
EMC_f is the equipment/item maintenance cost for the first year.
EIC is the equipment/item investment cost.
t is the equipment/item life expressed in years.
j is the amount by which maintenance cost increases annually.
i is the amount by which operating cost increases annually.

By differentiating Equation (6.7) with respect to t and then equating it to zero, we obtain the following expression:

$$t^* = \left[\frac{2EIC}{i+j}\right]^{1/2} \tag{6.8}$$

where

t^* is the equipment/item optimum-replacement interval.

Example 6.1

Assume that the following data are associated with a piece of medical equipment:

$$j = \$5,00$$
$$i = \$4,000$$
$$EIC = \$50,000$$

Calculate the optimum replacement interval for the equipment.

By inserting the given data values into Equation (6.8), we get:

$$t^* = \left[\frac{2(50,000)}{4000 + 5,00} \right]^{1/2}$$
$$= 4.71 \; years$$

Thus, the optimum replacement interval for the equipment in question is 4.71 years.

6.5.2 Model II

This model can be used to predict the number of spares required for a medical equipment/device in use. The number of spares required for each medical equipment/device in use is defined by the following equation:

$$\theta = \lambda t + z[\lambda t]^{1/2} \quad (6.9)$$

where

θ is the number of spares.
t is the mission time.
λ is the constant failure rate of the medical item under consideration.
z is associated with the cumulative normal distribution function.

The value of z depends on a specified confidence level for "no stock out." Thus, for a specified value of the confidence level, the value of z is obtained from the standardized cumulative normal distribution function table, given in various mathematical books. The standardized cumulative normal distribution function is expressed by:

$$P(z) = \frac{1}{2\sqrt{\Pi}} \int_{-\infty}^{z} e^{-\frac{x^2}{2}} dx \quad (6.10)$$

Example 6.2

Assume that the constant failure rate of a medical device is 0.0005 failures per hour. Determine the number of spare medical devices required during a 5,000-hour time period, if the confidence level for "no stock out" of the devices is 0.9032.

For the confidence level of 0.9032, from the standardized cumulative normal distribution function table, we obtain:

$$z = 1.3$$

Using the above value and the specified data in Equation (6.9) yields:

$$\theta = (0.0005)(5{,}000) + (1.3)\left[(0.0005)(5{,}000)\right]^{1/2}$$
$$\cong 5 \textit{ medical devices}$$

It means that five spare medical devices are needed for the specified period.

6.6 MEDICAL DEVICE-RELATED FAILURE DATA SOURCES

There are many sources for obtaining medical device-related failure data. Some of these are shown in Figure 6.2.

The Hospital Equipment Control System (HECS™) is a computerized system and it was developed by the Emergency Care Research Institute (ECRI) in 1985.[14,15] The system does the following:

- Provides effective support and analysis of clinical engineering-related operations through detailed work scheduling of items such as inspection, preventive maintenance, and repair
- Provides objective and good quality data on the basis of satisfactory statistical foundation

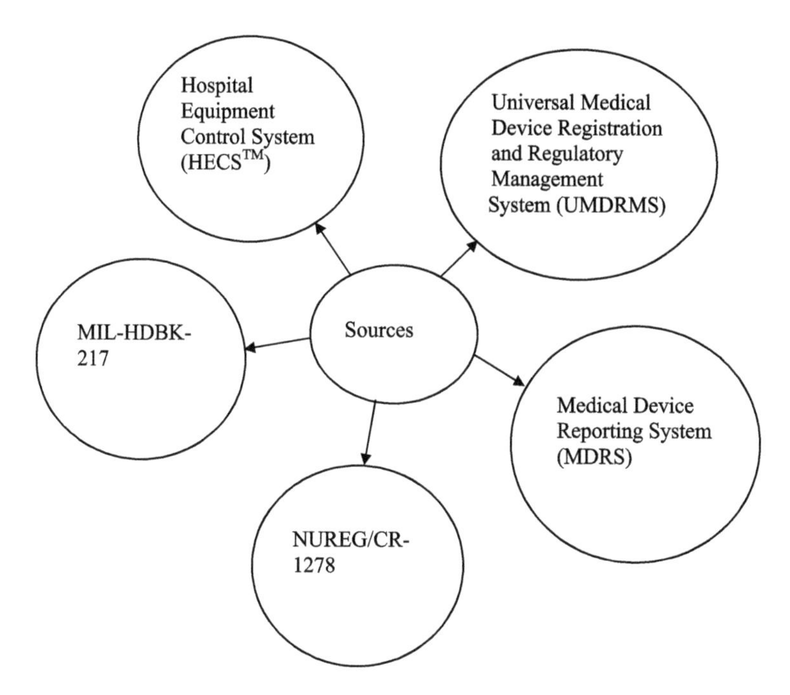

FIGURE 6.2 Sources for obtaining medical device-related failure data.

- Serves as the basic and effective building block of a new type of technology management system
- Provides cost-effective, economic, financial, and productivity data within the framework of each hospital

Furthermore, the system can also provide new information support, including data on the relative reliability of competing brands and models of clinical equipment, and reliable data on the cost of service, inspection, and preventive maintenance.

The Medical Device Reporting System (MDRS) is managed by the Center for Devices and Radiological Health, Food and Drug Administration (FDA).[16] It contains reports filed by device manufacturers concerning patient deaths and serious injuries allegedly involving their manufactured items, as well as failure of such items that may have caused serious injuries or deaths.

The Universal Medical Device Registration and Regulatory Management System (UMDRMS) was developed by ECRI and is designed to facilitate items such as tracking products for recall, safety, and reliability statistics and inventory control.[14,15]

MIL-HDBK-217 was developed by the United States Department of Defense and is widely used in the industrial sector to predict the failure rate of a given piece of equipment.[17] It contains generic failure rates for various parts, particularly the electronic ones, and a large number of mathematical models to predict failure rates of various types of parts in their actual use environment. This document can also be used to predict failure rates of medical equipment/devices or their parts.

The NUREG/CR-1278 handbook was developed by the United States Nuclear Regulatory Commission to perform human reliability analysis with emphasis on nuclear power plant applications.[18] The document contains failure data on various types of tasks performed by humans. The information contained in the document could be quite useful in analyzing medical devices in regard to the occurrence of human error.

6.7 ORGANIZATIONS FOR OBTAINING MEDICAL DEVICE-RELATED FAILURE DATA

There are many organizations that can be quite useful, directly or indirectly, in obtaining medical device-related failure data. Some of these organizations are listed below.

- Center for Devices and Radiological Health, Food and Drug Administration, 1390 Piccard Drive, Rockville, Maryland 20857
- Emergency Care Research Institute (ECRI), 5200 Butler Parkway, Plymouth Meeting, Pennsylvania 19462
- Parts Reliability Information Center (PRINCE), George C. Marshall Space Flight Center, National Aeronautics and Space Administration, Huntsville, Alabama 35812
- Reliability Analysis Center (RAC), Griffiss Air Force Base, U.S. Department of Defense, Rome, New York 13440
- National Technical Information Center (NTIS), U.S. Department of Commerce, 5285 Port Royal Road, Springfield, Virginia 22161

- Government-Industry Data Exchange Center (GIDEP), GIDEP Operations Center, Fleet Missile Systems, Analysis and Evaluation Group, U.S. Department of Defense, Corona, California 92878

6.8 MEDICAL DEVICE FAILURE-RELATED DATA

A wide variety of failure-related data, directly or indirectly, concerned with medical equipment/devices is available in the published literature. This section presents two types, directly or indirectly, of failure-related data: medical device specific and general component related.

The medical device specific data are concerned with the preproduction cause of a total of 1,143 medical device recalls for the period 1983 to 1987.[19] The causes for these recalls were grouped under seven distinct categories: device design, process design, component design, device software, label design, package design, and process software. The percentages of these causes were 33.8%, 27.8%, 23.6%, 7%, 4.3%, 3.3%, and 0.2%, respectively. It means that the highest cause for the medical device recalls was the device design (33.8%) and the lowest cause, the process software (0.2%).

The general component-related failure data are presented in Table 6.2 for certain items.[17,20] These items or components are often used in medical equipment/devices.

Table 6.2 Failure Data for Certain Items

Item (Part) Description	Failure Mode	Occurrence Probability	Failure Rate (Failures/10^6 Hours)
Push-button switch	Sticking	0.33	—
	Open	0.60	
	Short	0.07	
Neon lamps	—	—	0.20
Relay	Contact failure	0.90	—
	Coil failure	0.10	
Single-fiber optic connectors	—	—	0.10
Capacitor (ceramic)	Open circuit	0.01	—
	Short circuit	0.99	
Vibrators (MIL-V-95) (60-cycle)	—	—	15.00
Zener diode	Open circuit	0.5	—
	Short circuit	0.5	
Microwave ferrite devices: isolators and circulators (≤100 watts and use environment: ground, benign)	—	—	0.10
Micro-switch	High resistance	0.60	—
	Open circuit	0.30	
	No function	0.1	

6.9 PROBLEMS

1. Write an essay on medical device maintenance.
2. What are the six classifications under which medical equipment can be classified?
3. Define three medical equipment maintenance-specific indexes.
4. Describe the process considered useful in selecting a commercial computerized maintenance management system.
5. Assume that a medical equipment's investment cost is $100,000 and its operating and maintenance costs increase $5,000 and $2,000 per annum, respectively. Determine the optimum replacement interval for the equipment.
6. Prove Equation (6.8).
7. Discuss at least five useful sources for obtaining medical device-related failure data.
8. List at least six useful organizations for obtaining medical device-related failure data.
9. Define two broad maintenance indexes considered useful for application in health care organizations.
10. Discuss advantages and disadvantages of the index defined by Equation (6.6).

REFERENCES

1. Shooman, M.L. *Probabilistic Reliability: An Engineering Approach.* New York: McGraw-Hill, 1968.
2. Cohen, T. Validating Medical Equipment Repair and Maintenance Metrics: A Progress Report. *Biomedical Instrumentation and Technology* January/February (1997): 23–32.
3. Dhillon, B.S. *Medical Device Reliability and Associated Areas.* Boca Raton, FL: CRC Press, 2000.
4. Niebel, B.W. *Engineering Maintenance Management.* New York: Marcel Dekker, 1994.
5. Westerkamp, T.A. *Maintenance Manager's Standard Manual.* Paramus, NJ: Prentice Hall, 1997.
6. Stoneham, D. *The Maintenance Management and Technology Handbook.* Oxford, UK: Elsevier Science, 1998.
7. Hartmann, E., Knapp, D.J., Johnstone, J.J., Ward, K.G. *How to Manage Maintenance.* New York: American Management Association, 1994.
8. Dhillon, B.S. *Engineering Maintenance: A Modern Approach.* Boca Raton, FL: CRC Press, 2002.
9. Cohen, T. Computerized Maintenance Management Systems: How to Match Your Department's Needs with Commercially Available Products. *Journal of Clinical Engineering* 20 (1995): 457–461.
10. Ebel, G., Lang, A. Reliability Approach to the Spare Parts Problem. Proceedings of the Ninth National Symposium on Reliability and Quality Control, 1963, 85–92.
11. Wild, R. *Essentials of Production and Operations Management.* London: Holt, Rinehart and Winston, 1985.
12. Dhillon, B.S. *Mechanical Reliability: Theory, Models, and Applications.* Washington, DC: American Institute of Aeronautics and Astronautics, 1988.

13. Dhillon, B.S. *Engineering Maintainability*. Houston, TX: Gulf Publishing Company, 1999.
14. Emergency Care Research Institute (ECRI), 5200 Butler Parkway, Plymouth Meeting, PA 19462.
15. Nobel, J.J. Role of ECRI. In *The Medical Device Industry: Science, Technology, and Regulation in a Competitive Environment*, edited by N.F. Estrin, 177–198. New York: Marcel Dekker, 1990.
16. Arcarese, J.S. FDA's Role in Medical Device User Education. In *The Medical Device Industry: Science, Technology, and Regulation in a Competitive Environment*, edited by N.F. Estrin, 129–138. New York: Marcel Dekker, 1990.
17. Reliability Prediction of Electronic Equipment. Report MIL-HDBK-217. Washington, DC: U.S. Department of Defense.
18. Swain, A.D., Guttmann, H.E. Handbook of Human Reliability Analysis with Emphasis on Nuclear Power Plant Applications. Report NUREG/CR-1278. Washington, DC: U.S. Nuclear Regulatory Commission.
19. Grabarz, D.F., Cole, M.F. Developing a Recall Program. In *The Medical Device Industry: Science, Technology, and Regulation in a Competitive Environment*, edited by N.F. Estrin, 335–351. New York: Marcel Dekker, 1990.
20. Fries, R.C. *Reliable Design of Medical Devices*. New York: Marcel Dekker, 1997.

7 Human Error in Health Care

7.1 INTRODUCTION

Humans are an important element of the health care system and they are subjected to various types of errors. The history of the human error in health care may be traced back to 1848, when an error in the administering of anesthetic resulted in a death.[1,2] However, it was only in the late 1950s and the early 1960s when serious studies concerning medical errors were conducted.[3,4] These studies mainly focused on anesthesia-related deaths.

Today, human error in health care is the eighth leading cause of death in the United States and its cost is phenomenal. For example, the annual total national cost of medical adverse events is estimated to be around $38 billion, $17 billion of which is associated with preventable adverse events.[5] Furthermore, the annual cost of adverse drug events alone to a typical hospital facility is estimated to be approximately $5.6 million and $2.8 million of this figure is associated with preventable adverse drug events.[6]

Over the years a large number of publications on human error in health care have appeared. A comprehensive list of such publications covering the period 1963 to 2000 is available in ref. [7]. This chapter presents various different aspects of human error in health care.

7.2 HUMAN ERROR IN HEALTH CARE-RELATED FACTS, FIGURES, AND EXAMPLES

Some of the facts, figures, and examples, directly or indirectly, concerned with human error in health care are as follows:

- Each year approximately 100,000 Americans die due to human error in the health care system.[5]
- Deaths or serious injuries associated with medical devices, reported through the Center for Devices and Radiological Health (CDRH) of the U.S. Food and Drug Administration (FDA), accounted approximately 60% to user error.[8]
- More than 50% of the technical medical equipment-related problems are due to operator errors.[9]
- In 1993, a total of 7,391 people died as the result of medication errors in the United States.[7,10]
- The annual cost of medication errors is estimated to be approximately $7 billion in the United States.[11]

- An Australian study revealed that during the period 1988 to 1996, 2.4% to 3.6% of all hospital admissions were drug-related and 32% to 69% were preventable.[12]
- A study of 52 anesthesia-related deaths revealed that approximately 65% of the deaths were the result of human error.[4,13]
- In 1984, a study reviewed the records of 2.7 million patients discharged from New York hospitals and found that approximately 25% of the 98,609 patients who suffered an adverse event was the result of human error.[14]
- In 1993, a study revealed that over a 10-year period outpatient deaths due to medication errors increased by 8.48-fold in comparison with a 2.37-fold increase in inpatient deaths in the United States.[5,10]
- According to ref. [15], medication errors range from 5.3% to 20.6% of all administered doses.
- A study of 5,612 surgical admissions to a hospital identified 36 of them as adverse outcomes due to human error.[9,16]
- A study of 145 reports of adverse events involving patients in an intensive care unit of a hospital revealed 92 cases of human error and 53 cases of equipment/system failure.[17]
- A fatal radiation-overdose accident involving a Therac radiation therapy device was caused by a human error.[18]
- In 1990, a heart patient in a New York City hospital was mistakenly given 120 cubic centimeters per minute of a powerful drug instead of 12 cubic centimeters per hour. The patient died.[19]
- A patient died because of impeded airflow resulting from incorrect (i.e., upside-down) installation of an oxygen machine part.[20]

7.3 HUMAN ERROR IN MEDICATION

Past experiences indicate that the process of giving and taking medicine is not 100% error free. In fact, it is subject to error because the people involved, such as doctors, nurses, pharmacists, or even patients themselves, can make various types of mistakes. Although medication errors are considered unacceptable, they probably occur more often than they are actually reported due to reasons such as existence of poor or no error reporting system at all, loss of personal/organizational prestige, and loss of business/employment.

A medication error may be defined as any preventable event that may lead to incorrect medication use or patient harm while the medication is in the control of a health care professional, a consumer, or a patient.[21]

7.3.1 TYPES OF MEDICATION ERRORS

Medication errors may be classified under the following eight categories[21,22]:

- **Omission error.** This is the oversight to administer a recommended dose to a patient before the next scheduled dose (if any).

- **Incorrect dose error.** This is administering a dose to the patient that is higher or lower than the amount recommended by the legitimate prescriber.
- **Incorrect time error.** This is the administration of medication outside a predefined time interval from its scheduled administration time.
- **Incorrect drug preparation error.** This occurs when the drug product is incorrectly formulated/manipulated prior to its administration.
- **Prescribing error.** This is the incorrect drug selection, rate of administration, quantity, dose, dosage form, concentration, route, or instructions for use of a drug authorized by the legitimate prescriber. In addition, this type of error can also be illegible prescriptions or medication orders that lead to errors that reach the patient.
- **Unauthorized drug error.** This is administering medication to the patient that was not recommended by the patient's legitimate prescriber.
- **Incorrect dosage form error.** This is administering a drug product to the patient in a varying dosage form than recommended by the legitimate prescriber.
- **Incorrect administration method error.** This is the incorrect method or procedure employed in the administration of a drug.

7.3.2 Common Reasons for Occurrence of Medication Errors

There are many reasons for the occurrence of medication errors. Some of the common ones are shown in Figure 7.1.[21]

7.3.3 Useful General Guidelines to Reduce Occurrence of Medication Errors

Over the years various guidelines have been developed for use by medical professionals in reducing the occurrence of medication errors. Some of these guidelines are presented below.[7,21,23]

- Write legibly or utilize computer-generated prescriptions.
- Do not leave medications by the patient's bedside.
- Carefully check each patient's identification bracelet before administering a medication.
- Carry out dosage calculations on paper, not in head.
- Verbally educate all involved patients on the name and purpose of each medication.
- Carefully check the chart for all types of allergies when examining the medication administration record (MAR) against doctor's orders.
- Give proper attention to safety and efficacy in determining the amount of drug to be prescribed.
- Carefully check the drug label three times (i.e., at the time of removing the container from storage, before administering the drug, and prior to discarding or returning the container).

- Enquire from patients about allergies before administering any medication whatsoever. Aim to gain some knowledge of the patient's diagnosis to ensure the correctness of drug.
- Seriously examine the possibility of inadvertent drug substitutions when patients report some side effects.
- Avoid distractions when preparing medication for administration.

7.4 HUMAN ERROR IN ANESTHESIA

Anesthesiology may be described as an element of medicine that is specifically concerned with the processes of rendering patients insensible to various types of pain during surgery or when faced with chronic/acute pain states.[24] Although the first anesthetic-related death occurred in 1848, it took a long time to realize that human error plays an important role in the occurrence of anesthetic-related deaths.[1,2]

Human error in anesthesia may be defined in two ways: a mistake or a slip.[25] A mistake is a decision leading to an action or lack of action by the anesthesiologist or anesthetist that is causally linked to an actual or probable adverse outcome. Similarly, a slip is an action (or lack of action) by the anesthesiologist or anesthetist that did not occur according to the plan. Nonetheless, according to ref. [26], from 1952 to 1984, the risk of death due to anesthesia decreased from 1 in 2,680 to approximately 1 in 10,000.

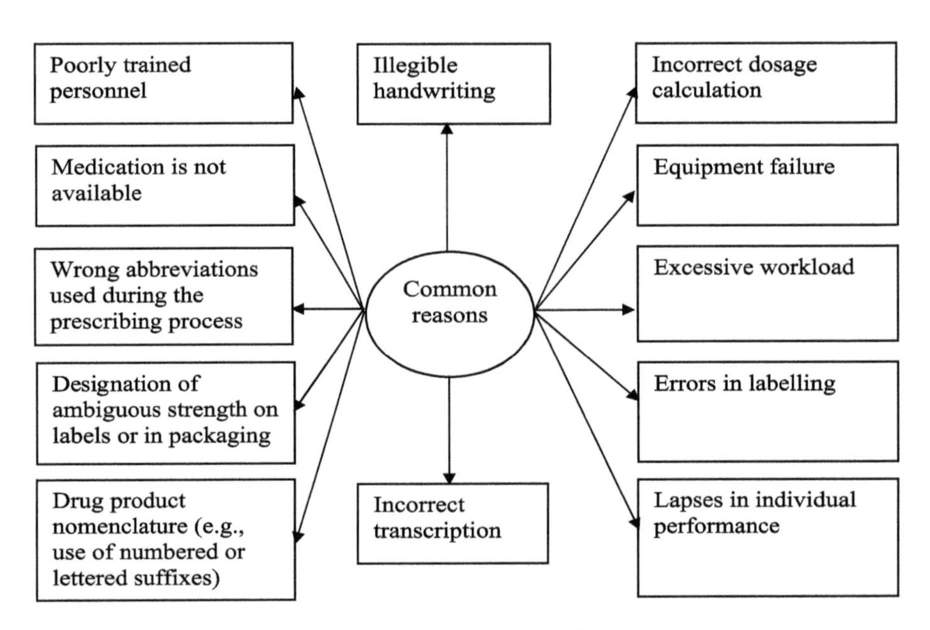

FIGURE 7.1 Common reasons for the occurrence of medication errors.

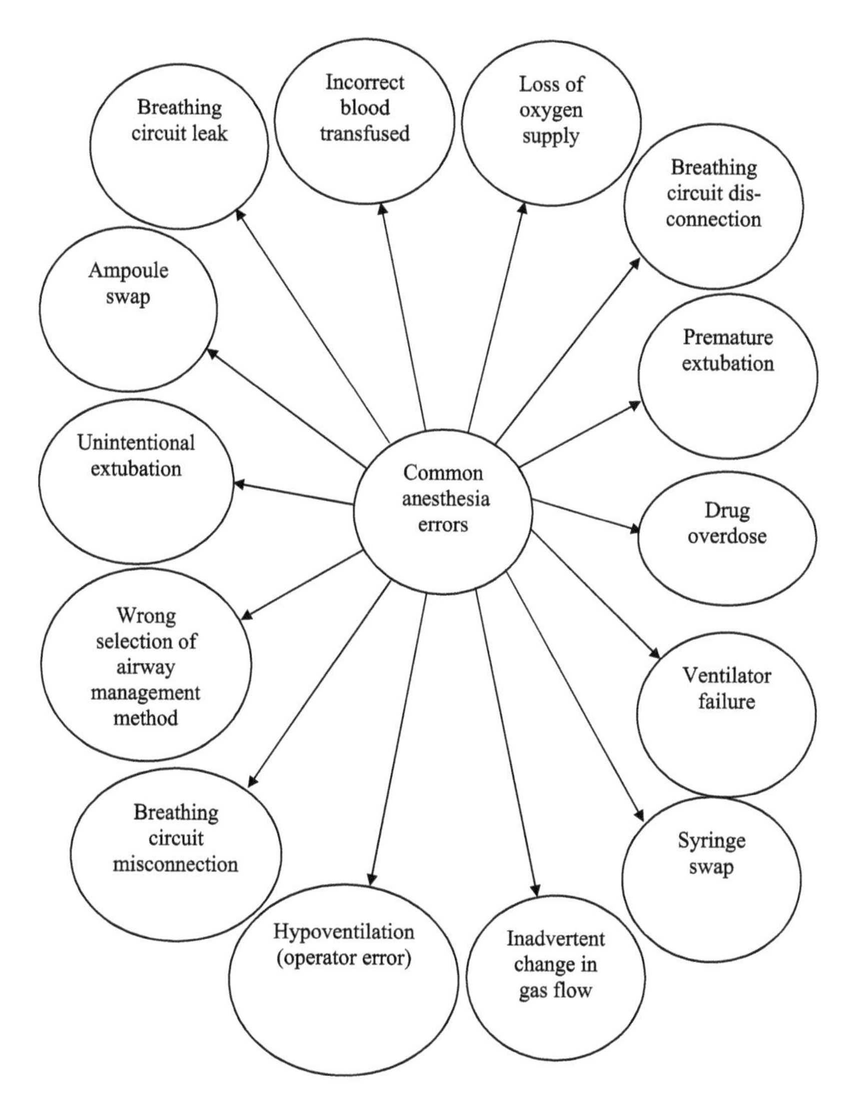

FIGURE 7.2 Common anesthesia errors.

7.4.1 Common Anesthesia Errors

Over the years, many studies have been performed to identify common anesthesia errors. Some of the commonly occurring anesthesia errors are shown in Figure 7.2.[27,28]

7.4.2 Common Causes of Anesthesia Errors

There are many causes for the occurrence of anesthesia errors. Some of the important ones are as follows[27,28]:

- Carelessness
- Poor familiarity with anesthetic method and surgical procedure
- Fatigue and haste
- Visual field restricted
- Inadequate anesthesia-related experience
- Emergency case
- Poor communication with laboratory personnel and surgical team
- Failure to carry out a proper checkout/history
- Inadequate familiarity with equipment
- Lack of skilled assistance or supervision
- Teaching activity in progress
- Excessive reliance on other personnel

7.4.3 Useful Methods to Prevent or Reduce Anesthetic Mishaps Due to Human Error

There are a number of useful methods that can be used to prevent or reduce the occurrence of anesthetic mishaps. Two of these methods are presented below.

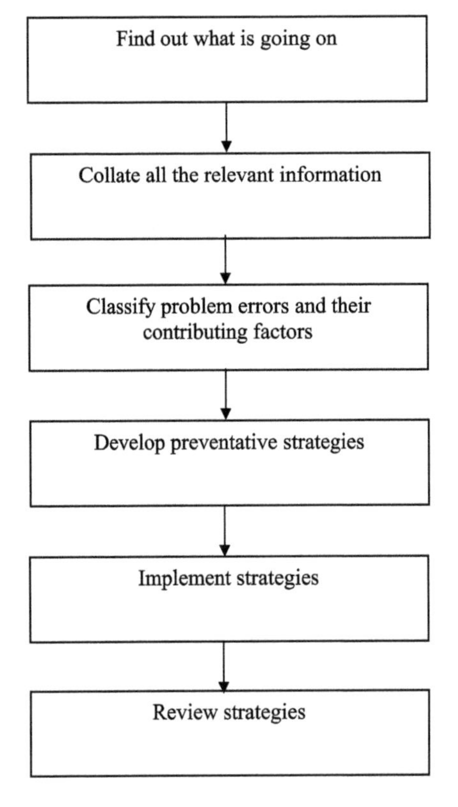

FIGURE 7.3 Steps for preventing or reducing the occurrence of anesthetic mishaps.

Method I

This method has proven to be quite useful in the past in preventing or reducing the occurrence of anesthetic mishaps due to human error. The method is composed of six steps, as shown in Figure 7.3.[29]

The step "Find out what is going on" is concerned with determining the existing state of affairs through the process of reviewing items such as morbidity committee reports, medical defense reports, incident monitoring studies, and mortality committee reports. The step "Collate all the relevant information" is concerned with collating information and gathering subsets for study and classification. This calls for the formation of a project team and having appropriate access to relevant information.

The step "Classify problem errors and their contributing factors" is basically concerned with the classification of problems, errors, and contributing factors under appropriate groups. The step "Develop preventive strategies" is concerned with preventing, avoiding, detecting, and minimizing the potential consequences; and ensuring that the strategies are made as practical as possible for their successful implementation.

The step "Implement strategies" is concerned with putting strategies into appropriate places. Effective communication is essential for the successful implementation of strategies. The final step "Review strategies" is basically concerned with assessing the effectiveness of strategies in action and taking appropriate corrective measures as necessary.

Method II

This method has proven to be quite effective to prevent errors and detect those errors and system failures that may slip unnoticed through the first line of defense. The method is basically composed of the following seven elements[7,25]:

- **Design and organize workspace.** This element calls for designing and organizing workspace by considering factors such as equipment, people, and tasks associated with anesthesia, so that the error occurrence probability is minimized and the accuracy and response speed are improved.
- **Assure equipment performance to effective levels.** This element calls for assuring the performance of anesthesia equipment by considering factors such as per use inspection, preventive maintenance, and recognition of obsolescence.
- **Supervise and train.** This element calls for assuring the availability of effective supervision and guidance as well as providing appropriate training in the area of technical skills, factual knowledge, or use of equipment and devices.
- **Use appropriate monitoring instrumentation and vigilance aids.** This element calls for using effective monitoring instrumentation and vigilance aids so that the occurrence of human error is reduced to a minimal level.
- **Develop and closely follow all appropriate preparation and inspection protocols.** This element calls for developing and following preparation and inspection protocols closely, because past experiences indicate that poor preparation for anesthesia or surgery was a contributory factor in at least 53% of fatal mishaps. Furthermore, the failure to perform an effective check of equipment used was identified as the most common factor associated with critical incidents.
- **Act on incident reports in an effective manner.** This element calls for taking appropriate corrective actions as per the findings of incident reports, in order to eradicate potential problems.
- **Recognize the limitations that influence individual performance.** This element calls for the recognition of factors that, directly or indirectly, influence individual performance. These factors include excessive haste, personal stress and life change events, and fatigue and the scheduling of work-rest cycles.[30,31]

7.5 HUMAN ERROR IN MEDICAL DEVICES

Human factors-related problems are often encountered in medical devices. Design-induced errors in the use of these devices can cause patient injuries and deaths. More specifically, misleading or illogical user interfaces can lead to errors even by the most experienced users. In fact, past experiences indicate that often-serious errors are committed by highly competent personnel.

Nonetheless, the likelihood of the occurrence of user errors increases quite significantly when medical devices are designed without giving appropriate attention to user cognitive, perceptual, and physical abilities. Various different aspects of human error in medical devices are presented below.

7.5.1 Medical Devices With High Incidence of Human Errors

Over the years, there have been many studies to identify medical devices with a high incidence of human error. Medical devices, in the order of most error-prone to least error-prone, are as follows[9,32]:

- Glucose meter
- Balloon catheter
- Orthodontic bracket aligner
- Administration kit for peritoneal dialysis
- Permanent pacemaker electrode
- Implantable spinal cord simulator
- Intravascular catheter
- Infusion pump
- Urological catheter
- Electrosurgical cutting and coagulation device
- Non-powered suction apparatus
- Mechanical/hydraulic impotence device
- Implantable pacemaker
- Peritoneal dialysate delivery system
- Catheter introducer
- Catheter guide wire
- Transluminal coronary angioplasty catheter
- External low-energy defibrillator
- Continuous ventilators (respirators)

7.5.2 Human Errors Causing User-Interface Design Problems

There are a number of user-interface medical device design-related problems that, directly or indirectly, cause user errors. Some of these problems are listed below.[33]

- Poorly designed labels
- Ambiguous or difficult to read displays
- Poor device design requiring unnecessarily complex installation and maintenance tasks

Table 7.1 Medical Device-Associated Operator Errors

Operator Error
1. Inadvertent or untimely activation of controls.
2. Incorrect decision making.
3. Incorrect selection of equipment/devices with regard to objectives and clinical requirements.
4. Incorrect interpretation of critical device outputs.
5. Incorrect improvisation.
6. Mistakes in setting equipment/device parameters.
7. Over-reliance on medical device/equipment automatic features, capabilities, or alarms.
8. Failure to follow prescribed instructions and procedures correctly.
9. Taking wrong actions in critical situations.

- Unnecessarily confusing or intrusive device alarms
- Complex or unconventional arrangements of items such as controls, displays, and tubing
- Confusing device operating instructions

7.5.3 Medical Device-Associated Operator Errors

There many operator-associated errors that occur during medical equipment/device operation. Some of these are presented in Table 7.1.[34]

7.5.4 Human Error Analysis Methods for Medical Devices

Over the years many methods and techniques have been proposed to perform various types of human error analysis. Some of these methods are as follows[7,9]:

- Failure modes and effect analysis (FMEA)
- Fault tree analysis (FTA)
- Markov method
- Force field analysis
- Barrier analysis

The first three of the above five methods are described in Chapter 3, and the fourth in Chapter 11. The last one is briefly described below.

Barrier Analysis

This method is based on the premise that an item possesses various types of energy (e.g., pharmaceutical reactions, mechanical impact, and heat) that can cause property damage and injuries. The method basically attempts to identify various types of energies associated with items and appropriate barriers to stop them from reaching humans or property.[35]

Although the method appears to be somewhat abstract, it can be an effective tool to identify serious hazards quickly. In the event an item has no designed-in barrier, the design personnel are alerted to incorporate one or more barriers as appropriate. It is to be noted that a barrier could be physical, behavioral, or procedural. Examples of physical and behavioral barriers are: protective gloves against blood-borne pathogens and drug interaction warnings because they influence users' behavior, respectively.

Some of the advantages of the barrier analysis are as follows:

- Easy to use and apply
- Easy to grasp
- Works quite well in combination with other methods
- Requires minimal resources

Similarly, some of the disadvantages of the barrier analysis are sometimes it promotes linear thinking and it can confuse causes and countermeasures. Additional information on the method is available in ref. [35].

7.6 HUMAN ERROR IN MISCELLANEOUS HEALTH CARE AREAS

Over the years, various studies have been performed to examine the occurrence of human error in some specific areas of health care, including emergency medicine, intensive care units, operating rooms, radiotherapy, and image interpretation. Human error in emergency medicine and intensive care units are discussed below, separately, and information on the other three areas is available in ref. [7].

7.6.1 Human Error in Emergency Medicine

There are around 100 million emergency department patient visits in the United States each year. Even a minute percentage of the human error occurrence during such visits can result in a substantial number of related adverse events.[36] Some of the facts and figures, directly or indirectly, concerned with emergency medicine are as follows:

- According to refs. [37, 38], more than 90% of the adverse events occurring in emergency departments are preventable.
- A study of missed diagnoses of acute cardiac ischemia in an emergency department reported that approximately 4.3% of the 1,817 patients with acute cardiac ischemia were incorrectly discharged from the emergency unit.[39]
- A study of the interpretation of radiographs reported that the rates of disagreement between emergency radiologists and physicians varied from 8% to 11%.[40]

The occurrence of human errors in emergency medicine can be reduced quite significantly by asking questions such as the following[39]:

- Is there any way to make the occurrence of human errors more visible?
- Will the presence of a pharmacologist be helpful in reducing human errors and adverse events in the emergency department setting?
- Are the computerized clinical information systems useful to reduce human errors and their related adverse events in the emergency department setting?
- Are there any effective change-of-shift approaches and lengths of shifts to reduce human error in emergency medicine?

7.6.2 Human Error in Intensive Care Units

Intensive care units are relatively a new phenomenon in hospitals in the United States. For example, by 1960 only 10% of the hospitals with more than 200 beds in the United States had intensive care units.[41,42] However, currently over 40,000 patients in intensive care units receive various types of services each day in the United States. The performance of such services is subject to various types of human errors. This is demonstrated by the facts and figures presented below.

- A study of critical incidents occurring in an intensive care unit during the period 1989 to 1999 reported that most of the incidents were due to staff personnel errors.[43]
- A study of incidents occurring in seven intensive care units over a period of 1 year reported that 66% of the incidents were the result of human factors-related problems.[44]
- A study concerned with the nature and causes of human errors committed in a six-bed intensive care unit reported 554 human errors during a 6-month period.[45] A further investigation of these errors revealed that 45% were committed by physicians and the remaining 55% by the nursing staff.

Over the years various factors for the occurrence of human errors in intensive care units have been identified. The important ones are as follows[43]:

- Poor communication
- Inadequate training and experience
- Staff shortage
- Night time

7.7 USEFUL GUIDELINES TO PREVENT OCCURRENCE OF HUMAN ERROR IN HEALTH CARE

Over the years various guidelines have been proposed to prevent the occurrence of human error in health care. Ten of these guidelines are as follows[46]:

- **Simplify.** This is concerned with reducing the number of steps in a work process, the nonessential procedures, equipment, and software; the number of times an instruction is given, etc.

- **Redesign the patient record for effectiveness.** This guideline calls for reviewing the effectiveness of the current record keeping form, as such a form could be too voluminous with buried important information.
- **Stratify.** This simply means to avoid over-standardization as much as possible because it can cause various types of errors.
- **Make improvements in communication patterns.** This is concerned with requiring team members in places such as intensive care units, operating rooms, and emergency departments to repeat orders to ensure that they have understood them clearly and correctly.
- **Automate cautiously.** This guideline calls for not over-automating because over-automation may prevent people such as operators from judging the true system state.
- **Respect human shortcomings.** This guideline calls for considering factors such as memory limitations, workload, stress, circadian rhythm, and time pressure in designing work systems and tasks, because various types of errors may occur, if factors such as these are overlooked during work system and task design.
- **Standardize.** This guideline calls for limiting unneeded variety in areas such as supplies, drugs, equipment, and rules, because as per past experiences when a procedure is carried out on a regular basis, the probability of doing it incorrectly is reduced quite significantly.
- **Use defaults effectively.** This guideline simply calls for making the correct action the easiest one. Nonetheless, a default may be described as a standard order or a rule that works quite well if nothing else intervenes.
- **Use affordances.** This is concerned with designing appropriate features in items that automatically force proper or correct use by providing appropriate clues to correct operation.
- **Use sensible checklists.** This is concerned with developing and using checklists sensibly and effectively.

7.8 PROBLEMS

1. Provide at least eight facts and figures, directly or indirectly, concerned with human error in health care.
2. What is a medication error?
3. What are the common causes for the occurrence of medication errors?
4. List at least 10 guidelines useful to reduce the occurrence of medication errors.
5. What are the common anesthesia errors?
6. What are the common causes for the occurrence of anesthesia errors?
7. List at least 15, in the order of most error-prone to least error-prone, medical devices with high incidence of human errors.
8. Discuss human error causing user-interface design problems.
9. Describe barrier analysis.
10. Discuss the occurrence of human error in the following two areas:
    - Emergency medicine

- Intensive care units

11. Discuss at least six most useful guidelines to prevent the occurrence of human error in health care at large.

REFERENCES

1. Beecher, H.K. The First Anesthesia Death and Some Remarks Suggested by It on the Fields of the Laboratory and the Clinic in the Appraisal of New Anesthetic Agents. *Anesthesiology* 2 (1941): 443–449.
2. Cooper, J.B., Newbower, R.S., Kitz, R.J. An Analysis of Major Errors and Equipment Failures in Anesthesia Management: Considerations for Prevention and Detection. *Anesthesiology* 60 (1984): 34–42.
3. Edwards, G., Morlon, H.J.V., Pask, E.A. Deaths Associated with Anaesthesia: A report on 1,000 Cases. *Anaesthesia* 11 (1956): 194–220.
4. Clifton, B.S., Hotten, W.I.T. Deaths Associated with Anaesthesia. *British Journal of Anaesthesia* 35 (1963): 250–259.
5. Kohn, L.T., Corrigan, J.M., Donaldson, M.S., eds. *To Err Is Human: Building a Safer Health System.* Washington, DC: Institute of Medicine, National Academy of Medicine, National Academies Press, 1999.
6. Bates, D.W., Spell, N.S., Cullen, D.J., et al. The Cost of Adverse Drug Events in Hospitalized Patients. *Journal of the American Medical Association* 277 (1997): 307–311.
7. Dhillon, B.S. *Human Reliability and Error in Medical System.* River Edge, NJ: World Scientific Publishing, 2003.
8. Bogner, M.S. Medical Devices: A New Frontier for Human Factors. *CSERIAC Gateway* IV (1993): 12–14.
9. Dhillon, B.S. *Medical Device Reliability and Associated Areas.* Boca Raton, FL: CRC Press, 2000.
10. Phillips, D.P., Christenfeld, N., Glynn, L.M. Increase in U.S. Medication-Error Deaths Between 1983 and 1993. *Lancet* 351 (1998): 643–644.
11. Wechsler, J. Manufacturers Challenged to Reduce Medication Errors. *Pharmaceutical Technology* February (2000): 14–22.
12. Roughead, E.E., Gilbert, A.L., Primrose, J.G., Sansom, L.N. Drug-Related Hospital Admissions: A Review of Australian Studies Published 1988–1996. *Medical Journal of Australia* 168 (1998): 405–408.
13. Cooper, J.B., Newbower, R.S., Long, C.D., et al. Preventable Anesthesia Mishaps. *Anaesthesiology* 49 (1978): 399–406.
14. Leape, L.L. The Preventability of Medical Injury. In *Human Error in Medicine*, edited by M.S. Bogner, 13–25. Hillsdale, NJ: Lawrence Erlbaum Associates, 1994.
15. Bindler, R., Bayne, T. Medication Calculation Ability of Registered Nurses. *Image—The Journal of Nursing Scholarship* 23 (1991): 221–224.
16. Couch, N.P., Tilney, N.L., Rayner, A.A., Moore, F.D. The High Cost of Low-Frequency Events. *New England Journal of Medicine* 304 (1981): 634–637.
17. Abranson, N.S., Wald, K.S., Grenvik, A.N.A., et al. Adverse Occurrences in Intensive Care Units. *Journal of the American Medical Association* 244 (1980): 1582–1584.
18. Casey, S. *Set Phasers on Stun: And Other True Tales of Design Technology and Human Error.* Santa Barbara, CA: Aegean Publishing, 1993.
19. Belkin, L. Human and Mechanical Failures Plague Medical Care. *New York Times*, March 31, 1992, B1, B6.
20. Swayer, D. *Do It By Design: An Introduction to Human Factors in Medical Devices.* Washington, DC: Center for Devices and Radiological Health, Food and Drug Administration, 1996.

21. Coleman, J.C. Medication Errors: Picking up the Pieces. *Drug Topics* March (1999): 83–92.
22. ASHP Guidelines on Preventing Medication Errors in Hospitals. *American Journal of Hospital Pharmacy* 50 (1993): 305–314.
23. Fox, G.N. Minimizing Prescribing Errors in Infants and Children. *American Family Physician* 53 (1996): 1319–1325.
24. Gaba, D.M. Human Error in Dynamic Medical Domains. In *Human Error in Medicine*, edited by M.S. Bogner, 197–223. Hillsdale, NJ: Lawrence Erlbaum Associates, 1994.
25. Gaba, D.M. Human Error in Anaesthetic Mishaps. *International Anaesthesiology Clinics* 22 (1984): 167–183.
26. Davies, J.M., Strunin, L. Anaesthesia in 1984: How Safe Is It? *Canadian Medical Association Journal* 131 (1984): 437–441.
27. Cooper, J.B. Toward Prevention of Anaesthetic Mishaps. *International Anaesthesiology Clinics* 22 (1984): 137–147.
28. Craig, J., Wilson, M.E. A Survey of Anaesthetic Misadventures. *Anaesthesia* 36 (1981): 933–938.
29. Runciman, W.B., Sellen, A., Webb, R.K., et al. Errors, Incidents, and Accidents in Anaesthetic Practice. *Anaesthesia and Intensive Care* 21 (1993): 506–518.
30. Selzer, M.L., Roges, J.E., Kern, S. Fatal Accidents: The Role of Psychopathology, Social Stress, and Acute Disturbance. *American Journal of Psychiatry* 124 (1968): 124–126.
31. Sheehan, D.V., O'Donnell, J., Fitzgerald, A., et al. Psychosocial Predictors of Accident/Error Rates in Nursing Students: A Prospective Study. *International Journal of Psychiatry Medicine* 11 (1981): 125–128.
32. Wiklund, M.E. *Medical Device Equipment Design*. Buffalo Grove, IL: Interpharm Press, 1995.
33. Rachlin, J.A. Human Factors and Medical Devices. *FDA User Facility Reporting: A Quarterly Bulletin to Assist Hospitals, Nursing Homes, and Other Device Users* 12 (1995): 86–89.
34. Hyman, W.A. Human Factors in Medical Devices. *Encyclopedia of Medical Devices and Instrumentation*, edited by J.G. Webster, 1542–1553. New York: John Wiley & Sons, 1988.
35. Maddox, M.E. Designing Medical Devices to Minimize Human Error. *Medical Device & Diagnostic Industry* 19, 5 (1997): 166–180.
36. Kyriacou, D.N., Coben, J.H. Errors in Emergency Medicine: Research Strategies. *Academic Emergency Medicine* 7 (2000): 1201–1203.
37. Wears, R.L., Leape, L.L. Human Error in Emergency Medicine. *Annals of Emergency Medicine* 34 (1999): 370–372.
38. Bogner, M.S., ed. *Human Error in Medicine*. Hillsdale, NJ: Lawrence Erlbaum Associates, 1994.
39. Pope, J.H., Aufderheide, T.P., Ruthazer, R., et al. Missed Diagnoses of Acute Cardiac Ischemia in the Emergency Department. *New England Journal of Medicine* 342 (2000): 1163–1170.
40. Espinosa, J.A., Nolan, T.W. Reducing Errors Made by Emergency Physicians in Interpreting Radiographs: Longitudinal Study. *British Medical Journal* 320 (2000): 737–740.
41. *Hospital Statistics: 1979 Edition*. Chicago: American Hospital Association, 1979.
42. Knaus, W.A., Wagner, D.P., Draper, E.A., et al. The Range of Intensive Care Services Today. *Journal of the American Medical Association* 246 (1981): 2711–2716.
43. Wright, D. Critical Incident Reporting in an Intensive Care Unit. Report, Western General Hospital, Edinburgh, Scotland, UK, 1999.

44. Beckmann, V, Baldwin, I., Hart, G.K., et al. The Australian Incident Monitoring Study in Intensive Care (AIMS-ICU): An Analysis of the First Year of Reporting. *Anaesthesia and Intensive Care* 24 (1996): 320–329.
45. Donchin, Y., Gopher, D., Olin, M., et al. A Look Into the Nature and Causes of Human Errors in the Intensive Care Unit. *Critical Care Medicine* 23 (1995): 294–300.
46. Crane, M. How Good Doctors Can Avoid Bad Errors. *Medical Economics* April (1997): 36–43.

8 Health Care Human Error Reporting Systems and Models for Predicting Human Reliability and Error in Health Care

8.1 INTRODUCTION

Past experiences indicate that the effectiveness of many decisions concerning human error in health care depends on the availability and quality of error data. It simply means that careful attention must be given when collecting and analyzing such data; otherwise these decisions may lead to costly problems. There are a number of human error reporting systems in use to collect health care human error-related data.[1-3] Some examples of the data, directly or indirectly, collected through such systems are medication-related incident rates and negligence rates for various clinical specialty groups.[3-5]

Mathematical models are frequently used in scientific fields to study various types of physical phenomena. Over the years a large number of mathematical models have been developed to study various aspects of human reliability and error in engineering systems.[6] Some of these models can also be used in the area of health care to study human reliability and error-related problems.

This chapter presents various different aspects of health care human error reporting systems and a number of mathematical models considered useful to tackle human reliability and error-related problems in health care.

8.2 KEY POINTS ASSOCIATED WITH CURRENT EVENT REPORTING SYSTEMS AND METHODS PRACTICED IN SUCH SYSTEMS

Currently there are a number of event reporting systems in use in the area of health care. They are subjected to factors (i.e., key points) such as the following[1]:

- Low occurrence of serious errors
- Appropriate resource adequacy
- Effectiveness in gathering error-related information from multiple sources
- Ability to detect unusual events

- Challenges such as getting sufficient participation in the programs and building an adequate response system
- Error monitoring by public/private systems
- Event reporters' perceptions and ability

Additional information on the above factors is available in refs [1, 6].

There are basically three methods practiced in current event reporting systems. They are as follows[1]:

- **Method I:** This method is mandatory internal reporting with audit. An example of the method is the Occupational Safety and Health Administration (OSHA) requiring organizations to keep all appropriate data internally according to the specified format and make it readily available during the onsite inspections by OSHA personnel.
- **Method II.** This method involves mandatory reporting to an outside agency. For example, the method is used by states such as Florida, California, and Ohio that require reporting by health care organizations for various accountability purposes.
- **Method III.** This method is voluntary and it reports confidentially to an outside agency to improve quality. The existing U.S. medication reporting programs fall under this category.

8.3 HEALTH CARE HUMAN ERROR REPORTING SYSTEMS

Over the years many, directly or indirectly, human error-related health care reporting systems have been developed. Some of these are as follows[1,3]:

- **Joint Commission on Accreditation of Healthcare Organizations (JCAHO) Event Reporting System.** This system was developed by the JCAHO in 1996 for hospitals to report sentinel events. A sentinel event is defined as an unexpected occurrence or variation that involves death/serious physical or psychological injury/the risk thereof.[1]
    - Whenever an organization experiences a sentinel event, JCAHO requires it to conduct root cause analysis (RCA) to identify the event causal factors. With regard to reporting of the event to the JCAHO, a hospital may voluntarily report an incident and submit the associated root cause analysis along with proposed suitable actions to carry out improvements.[1]
- **State Adverse Event Tracking Systems.** These systems are employed by various U.S. State governments to monitor the occurrence of adverse events in health care organizations. Two major problems, often cited to make greater use of the data reported in such systems, are lack of available resources and limitations in available data. Although these systems appear to provide a public response to investigate certain events, they are less likely to be effective in synthesizing information in communicating concerns to affected organizations or in analyzing where broad improvements might be necessary.

- **Medication Errors Reporting System.** This system was developed by the Institute for Safe Medication Practice (ISMP) in 1975 and is a voluntary medication error reporting system. The system receives reports from frontline practitioners and shares its information with pharmaceutical companies and the U.S. Food and Drug Administration (FDA). Currently, the system is operated by U.S. Pharmacopeia (USP).
- **Mandatory Internal Reporting with Audit System.** This is an OSHA requirement and it calls for keeping internal records of illness and injury by organizations employing more than 11 people. Although these organizations are expected to keep such records, they are not required to submit them routinely to OSHA. However, when they are selected in the annual OSHA survey of a sample of organizations, they must make these records available for onsite inspections.
- **Food and Drug Administration Surveillance System.** As part of this system, reports to the FDA on adverse events concerning medical products after their formal approval are submitted.[7] In the area of medical devices, the manufacturers report information on items such as deaths, failures, and serious injuries, and device users such as hospitals and nursing homes report deaths to both the FDA and manufacturers and serious injuries to manufacturers only.

With regard to drug-related suspected adverse events, reporting is voluntary to physicians, consumers, etc., but is mandatory for manufacturers.

8.4 MEDICAL AND GENERAL HUMAN ERROR-RELATED DATA

Human error data play a crucial role in making various types of decisions concerning human error. Over the years a significant amount of such data in both general and medical areas have been collected and analyzed.[6,8,9] Although most of these data are in the general area, some of it can also be used in the area of medicine.

Tables 8.1 and 8.2 present examples of the human error-related data available in medical and general areas, respectively.[5,9,10-13]

8.5 MODEL I: HUMAN RELIABILITY IN NORMAL WORK ENVIRONMENT

As most health care tasks are performed in normal environment, this mathematical model is concerned with predicting the reliability of health care professionals performing time-continuous tasks such as operating, monitoring, and tracking in general or normal environments. The reliability of health care professionals performing time-continuous tasks under normal environment can be estimated by using the following equation[6]:

$$R_{hp}(t) = e^{-\int_0^t \lambda_{hp}(t)\,dt} \tag{8.1}$$

Table 8.1 Examples of Human Error-Related Data Available in the Medical Area

Description	Rate (Annual Incidence/100 Beds)	Negligence Rate (%)
1.0. Incident		
1.1. Omission of dose	5	—
1.2. Incorrect medication administered	3	—
1.3. Medication administered to wrong person	1	—
1.4. Medication administered without physician order	1	—
1.5. Inappropriate dosing	7	—
2.0. Specialty group		
2.1. General surgery	—	28.0
2.2. Urology	—	19.4
2.3. Obstetrics	—	38.3
2.4. General medicine	—	30.9
2.5. Orthopedics	—	22.4

Table 8.2 Examples of Human Error-Related Data Available in the General Area

	Task/Error Description	Error Occurrence Probability
1.	Error in simple routine operation.	0.001
2.	General error of omission.	0.01
3.	Wrong switch (dissimilar in shape) selection.	0.001
4.	Read pressure gauge.	1.1 × 1/100
5.	Normal oral communication.	0.03
6.	Read flow/electrical meter.	1.4 × 1/100
7.	Error in a routine operation (where care is needed).	0.01
8.	Errors in simple arithmetic when self-checking as well.	0.03
9.	Remove drain tube.	1.9 × 1/1000
10.	Stressful complicated non-routine task.	0.3

where

$R_{hp}(t)$ is the health care professional's reliability at time t.

$\lambda_{hp}(t)$ is the time dependent failure rate or hazard rate of the health care professional.

The detailed derivation of Equation (8.1) is available in ref. [3]. This equation can be used to predict a health care professional's reliability when his/her time to human error follows any statistical distribution (e.g., gamma, Weibull, or exponential).

By integrating Equation (8.1) over the time interval [0, ∞], we obtain the following general expression for a health care professional's mean time to human error[6]:

$$MTTHE_{hp} = \int_0^{\infty} \left\{ e^{-\int_0^t \lambda_{hp}(t)\,dt} \right\} dt \tag{8.2}$$

where

$MTTHE_{hp}$ is the health care professional's mean time to human error.

Example 8.1

A health care professional is performing a certain task and his/her times to human error are described by Weibull distribution. Thus, the professional's time-dependent human error rate is expressed by:

$$\lambda_{hp}(t) = \frac{b\,t^{b-1}}{\theta^b} \tag{8.3}$$

where

t is time.
θ is the scale parameter.
b is the shape parameter.

Obtain an expression for the health care professional's reliability and calculate his/her reliability for an 8-hour mission when $b = 1$ and $\theta = 350$ hours (i.e., his/her mean time to human error).

By substituting Equation (8.3) into Equation (8.1) we get:

$$\begin{aligned} R_{hp}(t) &= e^{-\int_0^t \frac{b\,t^{b-1}}{\theta^b} dt} \\ &= e^{-\left(\frac{t}{\theta}\right)^b} \end{aligned} \tag{8.4}$$

By inserting the given data values into Equation (8.4), we obtain:

$$\begin{aligned} R_{hp}(8) &= e^{-\left(\frac{8}{350}\right)^1} \\ &= 0.9774 \end{aligned}$$

Thus, Equation (8.4) is the expression for the health care professional's reliability, and his/her reliability for the specified data values is 0.9774.

Example 8.2

Using Equations (8.2) and (8.3) obtain an expression for the health care professional's mean time to human error and for $b = 1$. Comment on the resulting expression.

By substituting Equation (8.3) into Equation (8.2), we get:

$$\begin{aligned} MTTHE_{hp} &= \int_0^\infty \left\{ e^{-\int_0^t \frac{b\, t^{b-1}}{\theta^b} dt} \right\} dt \\ &= \int_0^\infty e^{-\left(\frac{t}{\theta}\right)^b} dt \\ &= \theta \Gamma\left(1 + \frac{1}{b}\right) \end{aligned} \tag{8.5}$$

where

$\Gamma(\cdot)$ is the gamma function.

For $b = 1$, Equation (8.5) yields:

$$\begin{aligned} MTTHE_{hp} &= \theta \Gamma(2) \\ &= \theta \end{aligned}$$

Thus, for $b = 1$, θ is simply the health care professional's mean time to human error.

8.6 MODEL II: HUMAN RELIABILITY IN FLUCTUATING WORK ENVIRONMENT

This mathematical model is concerned with predicting a health care professional's reliability when performing his/her tasks in fluctuating environments (i.e., normal and stressful). Under this scenario, past experiences indicate that the occurrence of human errors may vary significantly from normal work environment to stressful work environment and vice-versa. Thus, this model is quite useful to calculate the professional's reliability and mean time to human error under the stated work conditions.

The model is subject to the following assumptions[14]:

- All human errors occur independently.
- Human error rates are constant.
- The rate of change of health care professional's work condition from normal to stressful and vice-versa is constant.

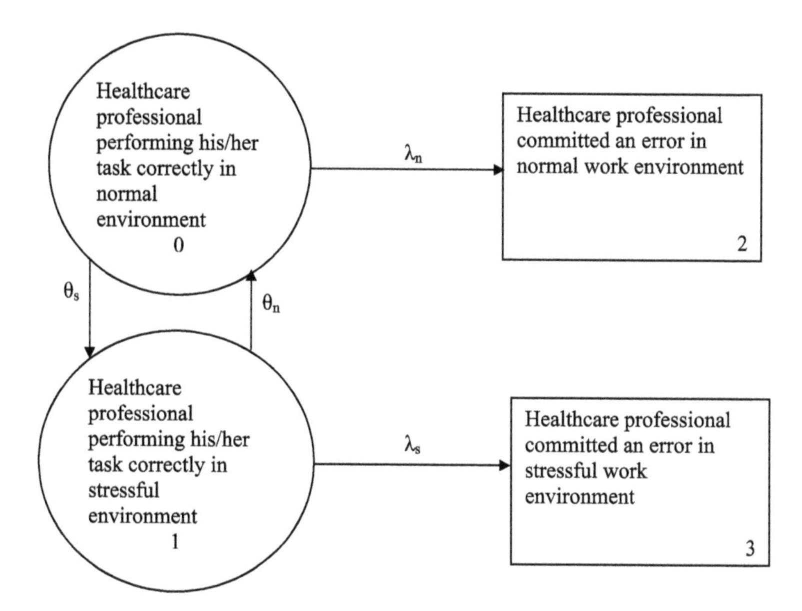

FIGURE 8.1 State space diagram for the health care professional working in normal and stressful environments.

The model state space diagram is shown in Figure 8.1. The numerals in boxes and circles denote health care professional's states. The following symbols are associated with the model:

- λ_n is the constant human error rate of the health care professional in normal work environment.
- θ_n is the constant transition rate from stressful work environment to normal work environment.
- λ_s is the constant human error rate of the health care professional in stressful work environment.
- θ_s is the constant transition rate from normal work environment to stressful work environment.
- j is the jth state of the health care professional; $j = 0$ means the health care professional performing his/her task correctly in normal environment, $j = 1$ means the health care professional performing his/her task correctly in stressful environment, $j = 2$ means the health care professional committed an error in normal work environment, $j = 3$ means the health care professional committed an error in stressful work environment.
- $P_j(t)$ is the probability of the health care professional being in state j at time t; for $j = 0, 1, 2, 3$.

By using the Markov method, we write down the following set of equations for the Figure 8.1 diagram[6,14]:

$$\frac{dP_0(t)}{dt} + (\lambda_n + \theta_s) P_0(t) = P_1(t)\,\theta_n \tag{8.6}$$

$$\frac{dP_1(t)}{dt} + (\lambda_s + \theta_n) P_1(t) = \theta_s\, P_0(t) \tag{8.7}$$

$$\frac{dP_2(t)}{dt} = P_0(t)\,\lambda_n \tag{8.8}$$

$$\frac{dP_3(t)}{dt} = \lambda_s\, P_1(t) \tag{8.9}$$

At time $t = 0$, $P_0(0) = 1$, and $P_1(0) = P_2(0) = P_3(0) = 0$.

Solving Equations (8.6) – (8.9) by using Laplace transforms, we obtain the following equations:

$$P_0(t) = (w_2 - w_1)^{-1}\left[(w_2 + \lambda_s + \theta_n)e^{w_2 t} - (w_1 + \lambda_s + \theta_n)e^{w_1 t}\right] \tag{8.10}$$

where

$$w_1 = \left[-C_1 + \sqrt{C_1^2 - 4C_2}\right]/2 \tag{8.11}$$

$$w_2 = \left[-C_1 - \sqrt{C_1^2 - 4C_2}\right]/2 \tag{8.12}$$

$$C_1 = \lambda_n + \lambda_s + \theta_n + \theta_s \tag{8.13}$$

$$C_2 = \lambda_n(\lambda_s + \theta_n) + \theta_s\,\lambda_s \tag{8.14}$$

$$P_2(t) = C_4 + C_5\,e^{w_2 t} - C_6\,e^{w_1 t} \tag{8.15}$$

where

$$C_3 = \frac{1}{w_2 - w_1} \tag{8.16}$$

$$C_4 = \lambda_n(\lambda_s + \theta_n)/w_1 w_2 \tag{8.17}$$

$$C_5 = C_3(\lambda_n + C_4 w_2) \tag{8.18}$$

$$C_6 = C_3(\lambda_n + C_4 w_2) \tag{8.19}$$

$$P_1(t) = \theta_s C_3 \left(e^{w_2 t} - e^{w_1 t} \right) \tag{8.20}$$

$$P_3(t) = C_7 \left[(1 + C_3) \left(w_1 e^{w_2 t} - w_2 e^{w_1 t} \right) \right] \tag{8.21}$$

$$C_7 = \lambda_s \, \theta_s \, / w_1 w_2 \tag{8.22}$$

The reliability of the health care professional is expressed by:

$$R_{hp}(t) = P_0(t) + P_1(t) \tag{8.23}$$

where

$R_{hp}(t)$ is the reliability of the health care professional at time t.

By integrating Equation (8.23) over the time interval [0, ∞], we obtain the following equation for the health care professional's mean time to human error[6,14]:

$$\begin{aligned} MTTHE_{hp} &= \int_0^\infty R_{hp}(t)\, dt \\ &= (\lambda_s + \theta_s + \theta_n) / C_2 \end{aligned} \tag{8.24}$$

where

$MTTHE_{hp}$ is the health care professional's mean time to human error.

Example 8.3

Assume that a health care professional is performing a certain task under normal and stressful environments. Constant transition rates from normal to stressful environment and vice-versa are 0.004 times per hour and 0.002 times per hour, respectively. The professional's constant error rates in normal and stressful environments are 0.001 errors per hour and 0.006 errors per hour, respectively. Calculate his/her mean time to human error.

By inserting the above specified data values into Equation (8.24), we get:

$$MTTHE_{hp} = \frac{0.006+0.004+0.002}{0.001(0.006+0.002)+(0.004)(0.006)}$$
$$= 375 \text{ hours}$$

Thus, the mean time to human error of the health care professional is 375 hours.

8.7 MODEL III: HUMAN RELIABILITY WITH CRITICAL AND NON-CRITICAL HUMAN ERRORS IN NORMAL WORK ENVIRONMENT

This mathematical model represents a health care professional performing a time-continuous task in normal work environment. He/she can make two types of errors: critical and non-critical. The model can be used to calculate the following items:

- The reliability of the health care professional at time t
- The probability of the health care professional making a critical human error at time t
- The probability of the health care professional making a non-critical human error at time t
- Mean time to human error of the health care professional

The state space diagram of the model is shown in Figure 8.2. The numerals in box and circles denote the health care professional's states and the other symbols used in the diagram are defined subsequently.

The following two assumptions are associated with the model:

- Critical and non-critical human error rates are constant.
- All human errors occur independently.

The following symbols are associated with the model:

- λ_1 is the health care professional's constant non-critical error rate.
- λ_2 is the health care professional's constant critical error rate.
- j is the jth state of the health care professional; $j = 0$ means the health care professional performing his/her task correctly, $j = 1$ means the health care professional committed a non-critical human error, $j = 2$ means the health care professional committed a critical human error.
- $P_j(t)$ is the probability of the health care professional being in state j at time t, for $j = 0, 1, 2$.

With the aid of the Markov method, we write down the following set of equations for the Figure 8.2 diagram[15]:

$$\frac{dP_0(t)}{dt} + (\lambda_1 + \lambda_2)P_0(t) = 0 \tag{8.25}$$

$$\frac{dP_1(t)}{dt} - \lambda_1 P_0(t) = 0 \tag{8.26}$$

$$\frac{dP_2(t)}{dt} - \lambda_2 P_0(t) = 0 \tag{8.27}$$

At time $t = 0$, $P_0(0) = 1$, $P_1(0) = 0$, and $P_2(0) = 0$.

Solving Equations (8.25) – (8.27) by using Laplace transforms, we get the following equations[15]:

$$P_0(t) = e^{-(\lambda_1 + \lambda_2)t} \tag{8.28}$$

$$P_1(t) = \frac{\lambda_1}{\lambda_1 + \lambda_2}\left[1 - e^{-(\lambda_1 + \lambda_2)t}\right] \tag{8.29}$$

$$P_2(t) = \frac{\lambda_2}{\lambda_1 + \lambda_2}\left[1 - e^{-(\lambda_1 + \lambda_2)t}\right] \tag{8.30}$$

The reliability of the health care professional at time t is given by

$$\begin{aligned} R_{hp}(t) &= P_0(t) \\ &= e^{-(\lambda_1 + \lambda_2)t} \end{aligned} \tag{8.31}$$

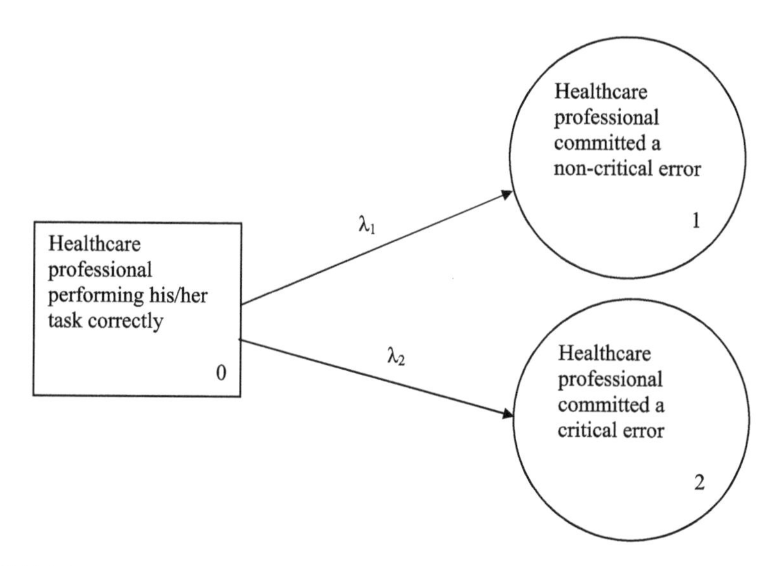

FIGURE 8.2 The model state space diagram.

where

$R_{hp}(t)$ is the reliability of the health care professional at time t.

By integrating Equation (8.31) over the time interval [0, ∞], we obtain the following equation for mean time to human error of the health care professional:

$$MTTHE_{hp} = \int_0^{\infty} e^{-(\lambda_1 + \lambda_2)t} dt = \frac{1}{\lambda_1 + \lambda_2} \tag{8.32}$$

where

$MTTHE_{hp}$ is the mean time to human error of the health care professional.

Example 8.4

A health care professional is performing a certain time-continuous task and his/her critical and non-critical error rates are 0.004 errors per hour and 0.01 errors per hour, respectively. Calculate his/her reliability for an 8-hour mission and mean time to human error.

By inserting the specified data values into Equation (8.31), we get:

$$R_{hp}(t) = e^{-(0.01 + 0.004)(8)} = 0.8940$$

Similarly, by substituting the given data values into Equation (8.32), we get:

$$MTTHE_{hp} = \frac{1}{(0.01 + 0.004)} = 71.43 \; hours$$

Thus, the health care professional's reliability and mean time to human error are 0.8940 and 71.43 hours, respectively.

8.8 PROBLEMS

1. Discuss three methods practiced in the current event reporting systems used in health care-related areas.
2. List at least five key points associated with current event reporting systems used in the area of health care.

3. Describe the following health care human error reporting systems:
    - Joint Commission on Accreditation of Healthcare Organizations Event Reporting System
    - Medication Errors Reporting System
    - Mandatory Internal Reporting with Audit System
4. Prove the end result of Equation (8.5).
5. Comment on Equation (8.1).
6. Prove Equations (8.10), (8.15), (8.20), and (8.21).
7. Prove that the sum of Equations (8.28), (8.29), and (8.30) is equal to unity.
8. Assume that a health care professional is performing a certain time-continuous task and his/her critical and non-critical error rates are 0.005 errors per hour and 0.02 errors per hour, respectively. Calculate his/her reliability for a 12-hour mission and mean time to human error.
8. Prove the end result of Equation (8.24).
9. Prove Equations (8.6) – (8.9).

REFERENCES

1. Kohn, L.T., Corrigan, J.M., Donaldson, M.S., eds. *To Err Is Human: Building a Safer Health System*. Washington, DC: Institute of Medicine, National Academy of Medicine, National Academies Press, 1999.
2. Billings, C.E. Some Hopes and Concerns Regarding Medical Event-Reporting Systems. *Archives of Pathology & Laboratory Medicine* 122 (1998): 214–215.
3. Dhillon, B.S. *Human Reliability and Error in Medical System*. River Edge, NJ: World Scientific Publishing, 2003.
4. Gurwitz, J.H., Samchez-Cross, M.T., Eckler, M.S., et al. The Epidemiology of Adverse and Unexpected Events in the Long-Term Care Setting. *Journal of the American Geriatric Society* 42 (1994): 33–38.
5. Brennan, T.A., Leape, L.L., Laird, N.M., et al. Incidence of Adverse Events and Negligence in Hospitalized Patients. *New England Journal of Medicine* 324 (1991): 370–376.
6. Dhillon, B.S. *Human Reliability: With Human Factors*. New York: Pergamon Press, 1986.
7. Washington, DC: U.S. Food and Drug Administration.
8. Dhillon, B.S. Human Error Data Banks. *Microelectronics and Reliability* 30 (1990): 963–971.
9. Gurwitz, J.H., Samchez-Cross, M.T., Eckler, M.S., et al. The Epidemiology of Adverse and Unexpected Events in the Long-Term Care Setting. *Journal of the American Geriatric Society* 42 (1994): 33–38.
10. Gertman, D.I., Blackman, H.S. *Human Reliability and Safety Analysis Data Handbook*. New York: John Wiley & Sons, 1994.
11. Kirwan, B. *A Guide to Practical Human Reliability Assessment*. London: Taylor & Francis Ltd., 1994.
12. Swain, A.D., Guttmann, H.E. Handbook of Human Reliability Analysis with Emphasis on Nuclear Power Plant Applications. Report NUREG/CR-1278. Washington, DC: U.S. Nuclear Regulatory Commission, 1983.
13. Irwin, I.A., Levitz, J.J., Freed, A.M. Human Reliability in the Performance of Maintenance. Report LRP 317/TDR-63-218. Sacramento, CA: Aerojet General Corporation, 1964.

14. Dhillon, B.S. Stochastic Models for Predicting Human Reliability. *Microelectronics and Reliability* 25 (1985): 729–752.
15. Dhillon, B.S. *Design Reliability: Fundamentals and Applications.* Boca Raton, FL: CRC Press, 1999.

9 Patient Safety

9.1 INTRODUCTION

Health care is considered a high-risk area because of adverse events arising from treatment rather than disease; and because it can lead to situations such as death, serious damage, complications, and patient suffering. Although many hospitals and health care settings have well-established procedures in place for ensuring patient safety, the health care area still lags behind other industrial sectors that have introduced various systematic safety processes. For example, the United States lost more American lives to patient safety-related problems every 6 months than it did in the entire Vietnam War period.[1] This also equates quite favorably to three fully loaded jumbo air planes crashing every other day for a period of 5 years.[1,2]

Needless to say, patient safety is a relatively recent initiative in the health care sector, emphasizing the reporting, analysis, and prevention of the occurrence of adverse health care events and medical error. The patient safety initiatives include many items including application of lessons learned from various industrial sectors, advancing technologies, and education of providers and the public. This chapter presents various important aspects of patient safety.

9.2 FACTS AND FIGURES

Some of the facts and figures, directly or indirectly, concerned with patient safety are as follows:

- A United Kingdom study, conducted by the Department of Health, revealed that adverse outcomes occur in approximately 10% of hospital admissions and cost the health service system an extra £2 billion annually in prolonged stays.[3,4]
- According to ref. [5], patient safety-related incidents cause harm in between 3% and 17% of hospital inpatients.
- According to a Harvard Medical Practice study, 3.7% of hospital admissions resulted in an adverse outcome.[6,7]
- A total of 18 patient safety indicators evaluated indicated approximately $9.3 billion excess charges and 32,591 deaths in the United States per year.[8]
- A study of 37 million hospitalizations in the Medicare population in the United States during the period 2000 to 2002 revealed that there were approximately 1.14 million patient safety incidents.[1]
- Patient safety-related incidents are more prevalent among medical admissions in comparison with surgical admissions.[1]
- A Canadian study revealed that out of approximately 2.5 million annual hospital admissions in Canada approximately 185,000 are associated with an adverse event and approximately 70,000 of these are potentially preventable.[9]

- In 1995, an Australian study concerning quality in health care in Australia discovered an adverse event rate of approximately 17% among hospital patients.[10]
- Each day approximately 150 people die in European Union countries due to a hospital-acquired infection.[11]
- According to ref. [11], 1 out of 10 hospitalized patients in European Union countries contracts a nosocomial (hospital-acquired) infection.
- According to ref. [10], at least 50% of medical equipment in most developing countries is not in usable condition, or partly in usable condition, at any given time.
- A study revealed that over a period of 10 years outpatient deaths due to medication-related errors in the United States increased by 8.48-fold in comparison with a 2.37-fold increase in inpatient deaths.[12]
- In the United States, 1 in every 136 hospital patients becomes quite ill because of acquiring an infection in the hospital.[13]
- According to ref. [13], roughly between 5% and 10% of patients admitted to hospitals in industrial countries acquire one or more infections.
- At any given point in time, more than 1.4 million individuals around the world are suffering from infections acquired in hospitals.[13]

9.3 PATIENT SAFETY GOALS

To improve patient safety in the United States, the Joint Commission on Accreditation of Healthcare Organizations (JCAHO) in 2001 established a number of patient safety goals with the aim of modifying them annually.[14,15] Some of its current goals are as follows[15]:

- **Enhance the accuracy of patient identification.** This goal calls for actions such as use at least two patient identifiers in providing treatment, care, or services and perform a final verification process to confirm the correct patient, procedure, and site before starting any invasive procedure.
- **Improve the safety of using all types of medications.** This goal calls for actions such as standardize and limit the number of drug concentrations used by the organization; label all types of medications, medication containers, or other solution on and off the sterile field; and identify and, at a minimum, annually review a list of look-alike/sound-alike drugs used by the organization, and take appropriate corrective measures to prevent errors involving the interchange of these drugs.
- **Enhance the effectiveness of communication among various caregivers**. This goal calls for actions such as listed below:
  - Standardize all symbols, acronyms, dose designations, and abbreviations that are not be used throughout the organizational setup.
  - In the case of verbal or telephone orders or for telephonic reporting of all critical test results, verify the entire order or test results by having the individual receiving the information properly record and "read-back" the entire order or test results.

    - Assess, measure, and take appropriate actions (as considered necessary) for improving the timeliness of reporting as well as the timeliness of receipt by the appropriate responsible licensed caregiver of critical test results and values.
- **Minimize the risk of patient harm resulting from falls.** This goals calls for the implementation of a fall reduction program, including an assessment of the program effectiveness.
- **Minimize the risk of health care-related infections.** This goal basically calls for complying with the current Centers for Disease Control and Prevention hand hygiene guidelines.
- **Minimize surgical fire risks.** This goal calls for establishing guidelines to minimize oxygen concentration under drapes and educating staff members, including operating licensed independent practitioners and anesthesia providers, on how to control heat sources and manage fuels with sufficient time for patient preparation.
- **Accurately and totally reconcile medications across the continuum of care.** This goal calls for actions such as the establishment of a process for comparing the patient's current medications with the ones ordered for the patient while under the care of the organization and communication of a complete list of the patient's current medications to the next provider of service when a patient is transferred or referred to another practitioner, setting, service, or level of care within or outside the current organization.
- **Encourage active involvement of patients in their own care as a patient safety strategy.** This goal calls for defining and communicating the means for patients and their family members to report concerns about safety and encouraging them to do so.

9.4 PATIENT SAFETY CULTURE AND ITS ASSESSMENT

Over the years various studies indicate that the existence of a positive safety culture or climate is absolutely essential to reduce the number of preventable patient injuries and their cost to society.[8,16-17] Also, there is an increasing tendency that it is essential to determine the proper relationship between the effects of safety culture on health care-related outcomes.[18,19] Two factors that hamper efforts in this direction are as follows:

- It is difficult to establish and validate patient safety outcomes across different patient populations and health care services.
- There is no single, accepted safety culture and climate model that identifies its components and their interrelationships.

If safety culture indeed plays a significant role in patient safety, then it is necessary to highlight which components of safety culture correlate with safety outcomes as well as to develop effective approaches for determining the existing type and nature of the safety culture and climate in individual wards, departments, and hospitals. The number of objectives that can be served by assessing the safety culture of an organization are shown in Figure 9.1.[16,20]

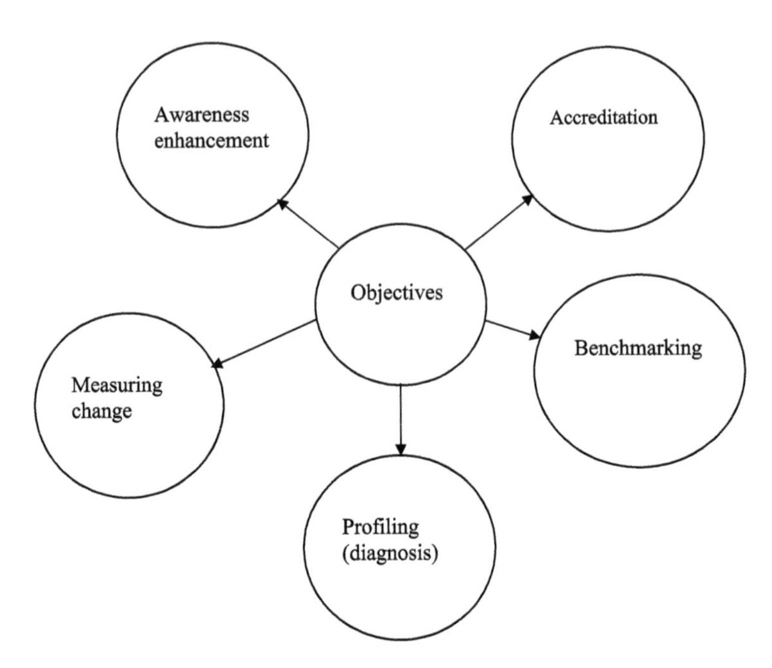

FIGURE 9.1 Objectives that can be fulfilled by assessing an organization's safety culture.

The objective—awareness enhancement—can serve to increase staff awareness, particularly when carried out in conjunction with other staff-oriented patient safety initiatives.

The objective—benchmarking—can be quite useful in evaluating the standing of a given unit in relation to a reference sample (i.e., comparable groups and organizations). The objective—profiling (diagnosis)—can be quite useful to determine a unit's specific safety culture or climate profile, including the identification of "weak" and "strong" points.

The objective—measuring change—can be quite useful in applying and repeating over a time period for detecting changes in attitudes and perceptions. Finally, the objective—accreditation—can serve as an element of a possibly mandated safety management review or accreditation program.

9.5 PATIENT SAFETY PROGRAM

In order to guide the planning and implementation of safety projects effectively, the existence of a comprehensive patient safety program is essential. An eight-step program for such a purpose is shown in Figure 9.2.[21] Step 1 is concerned with getting the feel for the norms or beliefs of the unit, team, or organization. The survey responses help to highlight perceptions of staff members with regard to how important they actually believe safety is to the unit and to the organization in question. Step 2 is concerned with educating staff members about the science of safety, so they better understand the reason that change is necessary and the importance of their participation.

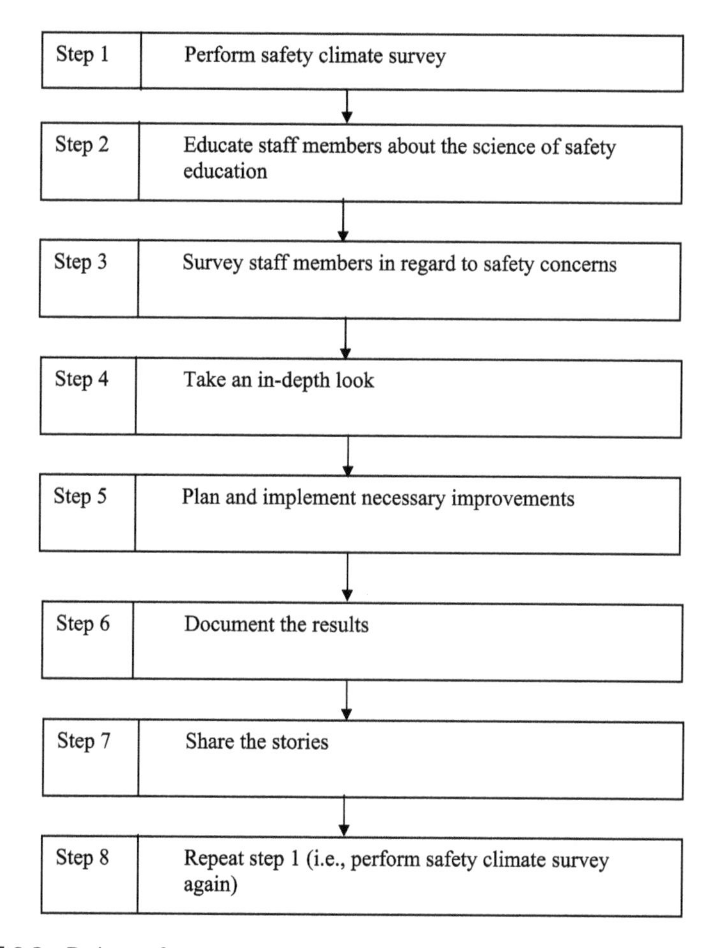

FIGURE 9.2 Patient safety program steps.

Step 3 is concerned with finding out the patient safety-related problems the staff members face on a daily basis. The problems for action are selected on the basis of factors such as the potential for harm, the frequency of occurrence, the resources needed to make the necessary change, and the likelihood of developing and implementing an intervention successfully. Step 4 is concerned with assembling an interdisciplinary team, once the specific safety initiative is identified, to take an in-depth look at the present system and processes of care.

Step 5 is concerned with planning and implementation of strategies to anticipate and prevent the occurrence of errors, or to minimize the potential for harm. The action plan incorporates a shared vision of goals that are essentially simple, measurable, and focused. Steps 6 and 7 are concerned with documenting the results and sharing stories with others about the progress and the obstacles encountered, respectively. The presentation of a balanced picture of the effort can be quite useful in spreading improvements to other areas and keeping the problem on the radar screen for the organization or unit in question. Finally, step 8 is concerned with repeating step 1 (i.e., perform safety climate survey) to assess if the organization or unit culture has shifted toward one that values safety as a top priority.

9.6 PATIENT SAFETY MEASURE SELECTION AND PATIENT SAFETY MEASURES AND ANALYSIS METHODS

A patient safety measure may be described as an instrument by which phenomena (e.g., outcomes, structures, or processes) associated with patient safety are quantified or translated into numbers.[22] A systematic approach is needed to select a patient safety measure. The approach involves assuring the clear identification of the problem, the evaluation of all possible options, and the selection of the measure on the basis of its ability to capture the phenomena of interest in an effective manner.

There are a number of sources from which patient safety problems may emerge. These include clinical experience (e.g., performance improvement, staff meetings, and sentinel events), professional sources (e.g., journals, standards, and networking at conferences), communications from external regulatory bodies, and feedback from patients. After the identification of the problem, the next major challenge is to select an appropriate measure. This involves finding an appropriate already established measure, adapting a measure to meet the need in question, or creating a new measure altogether. A six-step approach presented below can be used to select and develop patient safety measures.[22-25]

Step 1: Conduct a systematic literature review. This is basically concerned with determining items such as what reported data exist, the reported links between processes and outcomes, and the type of instruments that were used.

Step 2: Choose specific types of outcomes for evaluation. This is basically concerned with determining what outcome is important by keeping in mind that patient safety measures focus on structures and processes that are associated with patient injury or adverse health.

Step 3: Choose pilot measures. This is concerned with selecting pilot measures by determining the type of measure, evaluating the strength of evidence supporting the use of the measure, and assessing the feasibility of data collection.[25]

Step 4: Write design specifications for the measures. This is concerned with writing design specifications by defining the details of data collection, selecting a unit of analysis, defining the indicator, identifying the target population, etc.[25]

Step 5: Assess data validity and reliability. This is basically concerned with assessing the strength of the patient safety measure in the clinical environment, and establishing a plan for analysis.

Step 6: Pilot test the measures. This is concerned with pilot testing the patient safety measures with respect to factors such as reliability, precision, consistency, clarity, feasibility, validity, and utility.

Table 9.1 Important Clinical Patient Safety Performance Measures

	Performance Measure
1.	Number of deaths/serious injuries associated with a medical device.
2.	Number of sentinel events.
3.	Missing medication dose rate.
4.	Device-related intensive care unit bloodstream infection rate.
5.	Number of repeated sentinel events.
6.	Patient fall with injury rate.
7.	Nosocomial influenza rate.
8.	Reported significant medication errors.
9.	Methicillin-resistant *Staphylococcus aureus* bloodstream infection rate.

Some of the important patient safety performance measures are presented in Table 9.1.[14] These measures are number of deaths/serious injuries associated with a medical device, number of sentinel events, missing medication dose rate, device-related intensive care unit bloodstream infection rate, number of repeated sentinel events, patient fall with injury rate, nosocomial influenza rate, reported significant medication errors, and methicillin-resistant *Staphylococcus aureus* bloodstream infection rate.

Over the years, in the area of engineering safety, many methods have been developed to perform various types of safety analysis.[26,27] Some of these methods can also be used to perform patient safety-related analysis.[26-28] Five of these methods are briefly described below.[26,27]

- **Technic of operations review (TOR).** This method may simply be described as a hands-on type of analytical approach useful to identify the root causes of an operation failure.[28] It allows both management and employees to work together to analyze workplace accidents, incidents, and failures. The method is composed of a number of steps and is described in detail in refs. [26, 28, 29].
- **Root cause analysis (RCA).** This method was developed by the United States Department of Energy, and it may be described as a systematic investigation method that uses information gathered during an accident assessment to determine the underlying factors for the deficiencies that led to the occurrence of the accident.[30] RCA is performed by following a number of steps and starts with outlining the event sequence leading to an accident.
    - All in all, the performance of RCA helps to better understand the causal factors in the sequence of evolving events and the method is described in detail in refs. [27, 30, 31].
- **Hazard operability analysis (HAZOP).** This method was developed in the chemical industry and is used to identify hazards and operating problems throughout a facility. The method has proven to be an effective tool to identify unforeseen hazards designed into facilities and introduced into

existing facilities due to various factors, including changes made to operating procedures or process conditions.

- The method is composed of a number of steps and its main disadvantage is that it does not take into consideration the occurrence of human error in the final equation. Additional information on the method is available in refs. [29, 31].

- **Failure modes and effect analysis (FMEA).** This method was developed by the United States Department of Defense, and it may be described as an effective tool to perform analysis of each potential failure mode in the system to determine effects of such modes on the total system.[32] The method is widely used in the industrial sector and is composed of a number of steps.
  - It is described in detail in Chapter 3 and in refs. [33, 34].
- **Fault tree analysis (FTA).** This method was developed in the early 1960s for conducting safety analysis of the Minuteman Launch Control System.[35] Today, it is widely used in the industrial sector, particularly in nuclear power generation, to evaluate reliability and safety of engineering systems during their design and development.
  - A fault tree may be described as a logical representation of the relationship of primary events that may cause the occurrence of a specified undesirable event, called the "top event," and is depicted using a tree structure usually with OR and AND logic gates.
  - The method is described in detail in Chapter 3 and in refs. [34, 35].

9.7 PATIENT SAFETY ORGANIZATIONS

In recent years many organizations that advocate patient safety have been established in various parts of the world. Names and addresses of some of these organizations are presented in Table 9.2 for the benefit of readers.

9.8 PROBLEMS

1. Define patient safety.
2. List at least 10 facts and figures concerned with patient safety.
3. What are the typical goals of patient safety?
4. What are the objectives that can be fulfilled by assessing an organization's safety culture?
5. Describe a patient safety program useful to guide the planning and implementation of patient safety-related projects.
6. What is a patient safety measure?
7. Describe a method for selecting and developing patient safety measures.
8. List at least five clinical patient safety performance measures.
9. Compare RCA with FMEA.
10. List at least seven organizations that can be useful in obtaining patient safety-related information.

Table 9.2. Names and Addresses of Organizations Advocating Patient Safety

Name	Address
World Alliance for Patient Safety	c/o World Health Organization, 20 Avenue Appia CH-1211 Geneva 27 Switzerland
Safety and Quality in Health Care	Health Care Division, Department of Health, P. O. Box 8172 Stirling Street, Perth, Australia
National Patient Safety Agency	4–8 Maple Street, London W1T 5HD, United Kingdom
Agency for Healthcare Research and Quality	500 Gaither Road, Rockville, MD 20850
U.S. Food and Drug Administration	5600 Fishers Lane, Rockville, MD 20857-0001
Canadian Patient Safety Institute	100 – 1730 St. Laurent Blvd., Ottawa, Ontario K1G 5L1, Canada
National Patient Safety Foundation	132 Mass Moca Way, North Adams, MA 01247
Institute for Healthcare Improvement	20 University Road, 7th Floor, Cambridge, MA 02138
The Joint Commission on Accreditation of Healthcare Organizations (JCAHO)	601 13th Street NW, Suite 1150 N, Washington, DC 20005
Institute for Safe Medication Practices Canada	4711 Yonge Street, Suite 1600 Toronto, Ontario M2N 6K8, Canada
The National Quality Forum	601 13th Street NW, Suite 500 North, Washington, DC 20005
Institute for Safe Medication Practices	1800 Byberry Road, Suite 810, Huntingdon Valley, PA 19006
Australian Patient Safety Foundation	P.O. Box 400, Adelaide 5001, Australia

REFERENCES

1. Patient Safety in American Hospitals. Report, Health Grades, Inc., Golden, Colorado, July 2004.
2. Preventing Fatal Medical Errors. *New York Times*, December 1, 1999, A22.
3. Report of an Expert Group on Learning from Adverse Events in the National Health Service, Chaired by the Chief Medical Officer, Department of Health, London, 2000.
4. Hoyle, A. A Basic Guide to Patient Safety. *British Medical Journal/Careers* August (2005): 55–56.
5. Sary, A.B., Sheldon, T.A., Cracknell, A., Turnbull, A. Sensitivity of Routine System for Reporting Patient Safety Incidents in an NHS Hospital: Retrospective Patient Case Note Review. *British Medical Journal* 327 (2006): 432–436.
6. Brennan, T.A., Leape, L.L., Laird, N.M., et al. Incidence of Adverse Events and Negligence in Hospitalized Patients : Results of the Harvard Medical Practice Study. *New England Journal of Medicine* 324 (1991): 370–376.
7. Localio, A.R., Lawthers, A.G., Brennan, T.A., et al. Relation Between Practice and Claims and Adverse Events Due to Negligence: Results of the Harvard Medical Practice Study. *New England Journal of Medicine* 325 (1991): 245–251.
8. Zhan, C., Miller, M.R. Excess Length of Stay, Charges, and Mortality Attributable to Medical Injuries During Hospitalization. *Journal of the American Medical Association* 290 (2003): 1868–1874.

9. Baker, G.R., Norton, P.G., Flintoft, V., et al. The Canadian Adverse Events Study: The Incidence of Adverse Events Among Hospital Patients in Canada. *CMAJ: Canadian Medical Association Journal* 170 (2004): 1678–1686.
10. Patient Safety, Fact Sheets. World Health Professions Alliance, April 2002. www.whapa/factptsafety.htm.
11. Patient Safety, Health First Europe. Chaussee de Wavre 214D, 1050 Brussels, Belgium, 2007. www.healthfirsteurope.org/index.php?pid=82.
12. Phillips, D.P., Christenfeld, N., Glynn, L.M. Increase in US Medication-error Deaths Between 1983–1993. *Lancet* 351 (1998): 643–644.
13. Global Patient Safety Challenge: 2005–2006. Report, World Alliance for Patient Safety, World Health Organization, Geneva, Switzerland, 2005.
14. Poe, S.S. Using Performance Improvement to Support Patient Safety. In *Measuring Patient Safety*, edited by R. Newhouse, S.S., Poe, 13–25. Boston: Jones and Bartlett Publishers, 2005.
15. National Patient Safety Goals. The Joint Commission on Accreditation of Healthcare Organizations (JCAHO), 1 Renaissance Blvd., Oakbrook Terrace, Illinois, 2007. Also available online at www.jointcommission.org/patientsafety/Nationalpatientsafetygoals/07_npsg_facts.htm
16. Madsen, M.D., Andersen, H.B., Itoh, K. Assessing Safety Culture and Climate in Healthcare. In *Handbook of Human Factors and Ergonomics in Healthcare and Patient Safety*, edited by P. Carayon, 693–713. Mahwah, NJ: Lawrence Erlbaum Associates, 2007.
17. Kohn, L., Corrigan, J.M., Donaldson, M.S. *To Err is Human: Building a Safer Health System*. Washington, DC: Institute of Medicine, National Academy of Medicine, National Academies Press, 1999.
18. Gershon, R.R.M., Stone, P.W., Bakken, S., Larson, E. Measurement of Organizational Culture and Climate in Healthcare. *Journal of Nursing Administration* 34 (2004): 33–40.
19. Scott, T., Mannion, R., Marshall, M., Davies, H. Does Organizational Culture Influence Healthcare Performance? A Review of the Evidence. *Journal of Health Service Research Policy* 8 (2003): 105–117.
20. Nieva, V.F., Sorra, J. Safety Culture Assessment: A Tool for Improving Patient Safety in Healthcare Organizations. *Quality and Safety in Healthcare* 12 Supplement II (2003): 17–23.
21. Dawson, P.B. Moving Forward: Planning a Safety Project. In *Measuring Patient Safety*, edited by R.P Newhouse, S.S. Poe, 27–38. Boston: Jones and Bartlett Publishers, 2005.
22. Newhouse, R.P. The Metrics of Measuring Patient Safety. In *Measuring Patient Safety*, edited by R.P. Newhouse, S.S. Poe, 51–65. Boston: Jones and Bartlett Publishers, 2005.
23. McGlynn, E.A. Choosing and Evaluating Clinical Performance Measures. *Joint Commission Journal on Quality Improvement* 24 (1998): 470–479.
24. McGlynn, E.A. Selecting Common Measures of Quality and System Performance. *Medical Care* 41 (2003): 139–147.
25. Pronovost, P.J., Miller, M.R., Dorman, T., et al. Developing and Implementing Measures of Quality of Care in the Intensive Unit. *Current Opinion in Critical Care* 7 (2001): 297–303.
26. Dhillon, B.S. *Engineering Safety: Fundamentals, Techniques, and Applications*. River Edge, NJ: World Scientific Publishing, 2003.
27. Dhillon, B.S. Methods for Performing Human Reliability and Error Analysis in Health Care. *International Journal of Health Care Quality Assurance* 16 (2003): 306–317.

28. Hallock, R.G. Technique of Operations Review Analysis: Determine Cause of Accident/Incident. *Safety and Health* 60 (1991): 38–39.
29. Goetsch, D.L. *Occupational Safety and Health.* Englewood Cliffs, NJ: Prentice Hall, 1996.
30. Burke, A. Root Cause Analysis. Report, 2002. Available from Wild Iris Medical Education, P.O. Box 257, Comptche, California 95427.
31. Dhillon, B.S. *Human Error and Reliability in Medical System.* River Edge, NJ: World Scientific Publishing, 2003.
32. Omdahl, T.P., ed. *Reliability, Availability, and Maintainability (RAM) Dictionary.* Milwaukee, WI: American Society for Quality Control Press, 1988.
33. Palady, P. *Failure Modes and Effect Analysis.* West Palm Beach, FL: PT Publications, 1995.
34. Dhillon, B.S. *Design Reliability: Fundamentals and Applications.* Boca Raton, FL: CRC Press, 1999.
35. Dhillon, B.S., Singh, C. *Engineering Reliability: New Techniques and Applications.* New York: John Wiley & Sons, 1981.

10 Introduction to Quality in Health Care

10.1 INTRODUCTION

Many countries around the world spend a significant amount of their gross domestic product (GDP) on health care. For example, in 1992 the United States spent $840 billion on health care (i.e., 14% of its GDP).[1]

Although the history of quality in health care can be traced back to ancient times, in modern times, in the 1860s, the British nurse Florence Nightingale (1820–1910) helped to lay the foundation for modern health care quality assurance programs by advocating a pressing need for having a uniform system to collect and evaluate hospital statistics.[1] Her analysis of hospital statistics demonstrated that mortality rates varied quite significantly from one hospital to another. In 1910, a study by physician Abraham Flexner reported the poor quality of medical education in the United States. As the result of this study, 60 of 155 medical schools in the United States were closed by 1920.[2]

In 1914, physician E.A. Codman (1869–1940) studied the results of health care with respect to quality in the United States and emphasized various issues when examining the quality of care, including the accreditation of institutions, the importance of licensure or certification of providers, economic barriers to receiving care, the health and illness behavior of the patient, and the necessity of taking into consideration the severity or stage of the disease.[3]

Over the years, many other people have made advances to the field of quality in health care. A large number of references on the subject are listed at the end of this book. This chapter presents various introductory aspects of quality in health care.

10.2 REASONS FOR ESCALATING HEALTH CARE COSTS AND QUALITY DIMENSIONS OF TODAY'S HEALTH CARE BUSINESS

Over the years health care costs have been following an escalating trend. Some of the main reasons for this trend are as follows[4,5]:

- **The cost of poor quality.** This cost occurs when things are done incorrectly the first time. Some examples of the events that lead to such cost are retaking x-rays because the patient was not positioned correctly, the pharmacist could not read a prescription and calls the physician, and patient and physician wait in the clinic and delay treatment because the patient record cannot be found.

- **Variation in practice.** There is a wide variation in health care practice (i.e., from physician to physician, hospital to hospital, and state to state) due to a lack of standards. According to various studies, this variation, directly or indirectly, is an important factor in escalating health care costs.[4]
- **Aging population.** Today, 1 out of every 9 Americans is older than 65 years, and by 2010, this ratio is predicted to be 1 in 5. According to past experiences, usually people over 65 develop at least one chronic illness, thus affecting health care costs.[6,7]
- **New technology.** This may be defined as all forms of advancement in medical science due to the application of new drugs, equipment, or other forms of science. According to past experiences, usually much of the new technology supplements rather than replaces old methods, thus making it quite difficult to determine the effectiveness and cost benefits of the new procedures. All in all, as health care units attempt to use the new technology, tests overlap, are unnecessary, and increase health care costs.
- **Medical malpractice.** An increasing number of people in the United States file lawsuits against physicians, health care organizations, and other health care providers in the event of a question about the clinical decision, outcome, or procedure. As the result of such litigation, the cost of health care has been rising at a significant rate.
- **No incentive to control cost.** In the United States health care system, there is very little incentive to contain cost, as bills for care are usually being paid by third parties (i.e., not by those receiving services); thus, patients are not price sensitive. This, in turn, drives up health care costs.

There are many quality dimensions of today's health care business. Some of these are as follows[8]:

- Health care customers such as patients, relatives, and physicians appear to be quite sensitive to quality differences; consequently, their choice of facilities will be affected accordingly.
- Quality is not a departmental activity; it is the responsibility of all concerned people, such as physicians, nurses, biomedical technicians, and pharmacists.
- As patient satisfaction is affected by competitive offerings, it should be assessed at the time of the health care facility (hospital) visit and again after discharge.
- As high quality is expected to generate greater customer loyalty, over a time period, it should translate into a greater rate of return visits and referrals to health care facilities.
- The services offered by other competitive health care facilities should be studied to determine what customers really mean when they say one health care facility or service is better than the other with regard to quality.
- There is indisputably a connection between quality and productivity.
- If profitability and quality are interlinked closely, health care facilities must aim to satisfy and exceed the quality level of their immediate competitors.

- Continuous improvement should become the deliberate goal rather than achieving simply an acceptable level of quality on the basis of some internal standards or Joint Commission on Accreditation of Healthcare Organizations (JCAHO) requirements.
- Conformance to internal and JCAHO standards appears to have become a secondary issue, to be pursued only after effectively defining the requirements of the health care facility's (e.g., hospital's) customers.
- It is not good enough to do all the right things well; health care facilities also have to do them much better with regard to their competitors.

10.3 HEALTH CARE QUALITY GOALS AND THEIR ASSOCIATED STRATEGIES

Although there could be many goals of quality in health care, the four important ones are as follows[9]:

Goal I: Provide good person-centered compassionate care that respects individual dignity as well as is responsive to the needs of people such as patients, residents, and families.
Goal II: Establish an effective system perspective on communicating and analyzing data and information concerning quality, appropriateness, outcomes, and cost of care.
Goal III: Support a quality management mechanism that is effective to further coordination of care across the continuum of providers and services.
Goal IV: Involve all employees, physicians, and board members in system efforts to implement total quality management principles.

Two useful strategies associated with goal I are as follows[9]:

- Ensure the assessment of satisfaction of people such as patients, employees, and medical staff on a regular basis, by incorporating survey standards and benchmarking.
- Aim to maximize patients and families involvement in the care experience by improving patient involvement in care choices as well as by the use of shared decision making.

Four useful strategies associated with goal II are as follows[9]:

- Establish a system plan for addressing information needs associated with quality management, including items such as a core clinical data set, improved analysis of available information, and common definitions.
- Improve the skills and competencies of system manpower associated with quality management.
- Examine the implications of new developments with regard to electronic medical records.

- Document and share important quality performance and outcome-related studies throughout the total system.

Two useful strategies associated with goal III are as follows[9]:

- Evaluate how the development of integrated delivery systems can be useful in promoting access and quality of care.
- Improve and apply case management models, particularly across the continuum of all related services.

Three useful strategies associated with goal IV are as follows[9]:

- Develop appropriate educational programs for people such as board members, physicians, and employees.
- Develop and apply appropriate management models that help to promote teamwork and participatory decision making.
- Actively involve all concerned physicians in developing treatment protocols and improving care systems.

10.4 HEALTH CARE AND INDUSTRIAL QUALITY COMPARISONS AND QUALITY IMPROVEMENT VERSUS QUALITY ASSURANCE IN HEALTH CARE INSTITUTIONS

By comparing quality from an industrial perspective with the quality from a health care perspective, it can be concluded that both are surprisingly similar and at the same time both have strengths and weaknesses.[10] Some limitations of the industrial model with respect to health care are as follows[10]:

- Totally overlooks the complexities of the patient–practitioner relationship
- Emphasizes greater attention to supportive activities and less to clinical-related ones
- Downplays the practioner's knowledge, skill, and motivation
- Gives less emphasis on influencing professional performance through "education, supervision, encouragement, retraining, and censure"

Some useful points that the professional health care model can learn from the industrial model are as follows[10]:

- The need for greater emphasis on items such as consumer requirements, values, and expectations
- The need for a greater role by management personnel to assure clinical care quality
- The need to extend the physicians' self-governing and self-monitoring tradition to other people within the organization

Table 10.1 Comparisons of Quality Assurance and Quality Improvement in Health Care Organizations

	Area (Characteristic)	Quality Assurance	Quality Improvement
1.	Focus	Physician	Processes
2.	Goal	Regulatory compliance	Satisfy customer requirements
3.	Customers	Regulators	Patients, caregivers, technicians, payers, families, etc.
4.	Functions involved	Few (mainly physicians)	Many (clinician and support system)
5.	Objective	Control	Breakthrough
6.	Action taken	Recommend improvements	Implement useful improvements
7.	Performance measure	Outside/external standards	Need/capability
8.	Direction	Committee members or coordinator	Decentralized through the management line of authority
9.	Defects investigated	Special causes/outliers	Common and special causes
10.	Tampering	Common	Quite rare
11.	Review approach	Summary/abstract	Analysis
12.	Participants	Peers	Everyone

- The need to develop applications of statistical quality control approaches to the health care monitoring area
- The need for greater emphasis on the design of processes and systems as a means of assuring quality
- The need for greater attention to education and training in quality assurance and monitoring for all concerned individuals

In the past many authors have pointed out a number of important differences between quality assurance and quality improvement in health care institutions.[11-15] A clear understanding of such differences is necessary to promote quality assurance in health care organizations. Many of these differences are presented in Table 10.1. [11-15]

10.5 COMMON HEALTH CARE ADMINISTRATORS' APPEALS FOR TOTAL QUALITY AND PHYSICIANS' REACTIONS

From time to time health care administrators appeal for total quality based on experience of previous efforts. Some of their common appeals and corresponding reactions of physicians are presented in Table 10.2.[16]

10.6 STEPS FOR IMPROVING QUALITY IN HEALTH CARE

A 10-step approach presented below can be used to improve quality in health care organizations.[17] The material presented in parentheses demonstrates "how TQM can help" in each of these steps.

Step 1: Assign responsibility. Provides appropriate reports to the board of trustees, management, and medical staff members to monitor and review clinical-related activities.

Step 2: Delineate scope of care. Evaluates clinical-related performance by service.

Step 3: Highlight all main aspects of care. Highlights pertinent activities, functions, outcomes, and treatments.

Step 4: Identify all appropriate indicators. Incorporates a benchmarking method of other similar health care facilities for establishing an appropriate objective and measurable indicators.

Step 5: Develop appropriate thresholds for evaluation. Develops statistically significant thresholds by comparing appropriate indicators across customized or national norms through the benchmarking process.

Step 6: Collect and organize all relevant data. Uses available data and organizes it into the form of standard reports.

Table 10.2 Common Appeals of Health Care Administrators for Total Quality Management (TQM) and Corresponding Physicians' Reactions

Health Care Administrators' Appeal	Physicians' Reaction
1. TQM is quite different from the traditional quality assurance (QA) program because it focuses on improving all concerned processes rather than a "bad apple" approach.	TQM is basically QA in different clothing.
2. TQM makes use of the scientific approach for improving processes.	This is nothing new. (We have always used the scientific approach in the past.)
3. TQM will be useful to physicians for taking a proactive role in affecting changes that considerably influence clinical care.	The application of the TQM concept will result in additional meetings for already time-constrained physicians.
4. The TQM concept places emphasis on examining and improving systems rather than focusing on outliers.	The TQM concept is applicable to industrial processes and administrative systems, but not to the clinical care of individual patients.
5. Multidisciplinary teams of all individuals involved in the process are the real foundation of TQM efforts.	The application of the TQM concept will wrest control of the patient care process from physicians.
6. The TQM concept is a useful structured process for reducing duplication, rework, and inappropriate utilization.	The TQM concept is another cost-cutting tool by management that will limit access to resources physicians need for their patients.
7. The TQM concept indicates that in controlled systems, patient care can be standardized effectively, thus generates better quality and efficiency by reducing variation.	The application of the TQM concept is a further encroachment on the physician–patient relationship, as it is impossible to standardize patient care like industrial processes.

Step 7: Assess care when relevant thresholds are reached satisfactorily. Highlights appropriate statistically meaningful areas of exemplary performance and opportunities for the purpose of improving quality and provides individual patient listings of the purpose of focused reviews.

Step 8: Initiate appropriate measures to improve care. Highlights appropriate care processes for corrective actions.

Step 9: Determine effectiveness and maintain the gain. Continuously monitors all relevant indicators for documenting changes in performance over a time period.

Step 10: Communicate results to all concerned people. Provides relevant executive, trend, and standardized reports for clear examples of effective quality-related improvements and helps to communicate important issues.

10.7 IMPLEMENTATION OF SIX SIGMA METHODOLOGY IN HOSPITALS AND ITS ADVANTAGES AND IMPLEMENTATION BARRIERS

Although the history of Six Sigma as a measurement standard may be traced back to the father of the normal distribution, Carl F. Gauss (1777–1855), it was the Motorola company in the 1980s that explored this standard and created the methodology and necessary cultural change associated with it.

Six Sigma may be described as a methodology implementation directed at a measurement-based strategy that develops process-related improvements and varied cost reductions throughout a company or an organizational set up. In some companies, Six Sigma simply means an effective measure of quality that strives for near perfection.

In recent years, some health care organizations have started to apply the Six Sigma methodology to their operations. In the implementation of define, measure, analyze, improve, and control (DMAIC) Six Sigma methodology in industrial organizations, basically a total of nine steps, listed below, are involved.[18]

1. Provide training and start project.
2. Identify all stakeholders and gather data.
3. Map and analyze processes including pertinent processes.
4. Identify appropriate metrics for each process and set targets for improvement.
5. Estimate costs associated with defects and develop recommended alternative solutions.
6. Choose/implement appropriate solutions.
7. Rework the process.
8. Review sustainability of process improvements.
9. Estimate performance improvement as applicable.

The above steps can be tailored accordingly for the implementation of the methodology in health care organizations.

There are many advantages of implementation of Six Sigma methodology in hospitals or in other health care organizations. Some of the important ones are as follows[18]:

- Measurement of essential health care performance requirements on the basis of commonly used standards (i.e., through the use of statistical analysis and hypothesis testing)
- The implementation of the methodology with emphasis on improving customers' lives could result in the involvement of more health care people, such as professionals and support personnel, in the quality improvement process
- Better job satisfaction of health care workers
- Establishment of shared accountability in regard to continuous quality improvement

There are many potential barriers to the implementation of Six Sigma programs in health care organizations. For example, some of these barriers in regard to hospitals are as follows[18]:

- Costs (e.g., start-up and maintenance)
- Nursing shortage
- Poor support from physicians
- Governmental regulations
- Difficulty in getting baseline data on process performance
- Risk of the methodology being implemented to only easily measurable non-patient care processes
- Rather long project ramp-up times (i.e., 6 or more months)

10.8 QUALITY INDICATORS FOR USE IN HOSPITAL DEPARTMENTS

There are a large number of quality-related indicators used in health care. Some of the statements that can be made regarding such indicators are as follows[19]:

- An indicator is always a number that tells something about quality.
- Indicators are signals, flags, or signs.
- Indicators are not infallible and are of two types: sentinel events (i.e., single, highly significant events such as a fire, a death, or a lawsuit) and rate-based indicators (e.g., rate of infections and rate of accidents).
- Indicators can be defined as positive or negative and in five dimensions.

There are many reasons for using indicators. Some of these are provide a highly efficient means of monitoring performance, save time, good indicators are widely comparable, and they can be the source of further investigations.[19] Nonetheless, a good indicator should satisfy the criteria in the order listed: significance, validity, data availability, sensitivity, and professional respect.[19]

A number of quality indicators for use in various hospital departments are presented below.[19]

10.8.1 Department: Nursing, Acute Care

Some of the quality or performance indicators (along with their corresponding principal functions in parentheses) are as follows:

- Failure to assess correctly (assessment and planning of care)
- Complaints from family (personal care)
- Infection rate (treatment/intervention)
- Medical incidents (treatment/intervention)
- Number of discrepancies on safety inspections (provision of safe and comfortable environment)

10.8.2 Department: Physiotherapy

Some of the useful quality or performance indicators (along with their corresponding principal functions in parentheses) are as follows:

- Percentage of patients achieving treatment goals on time (treatment/inpatient)
- Percentage of appropriate/inappropriate referrals (communication and charting)
- Lengths of time on waiting list (treatment/inpatient)

10.8.3 Department: Social Work Services

Some of the quality or performance indicators (along with their corresponding principal functions in parentheses) are as follows:

- Percentage of patients seen by social work services (assessment/intake)
- Discharge delays (community liaison/referral)
- Timeliness of service (patient/family counseling)
- Readmissions with social complications (patient/family counseling)

10.8.4 Department: Pharmacy

Some of the useful quality or performance indicators (along with their corresponding principal functions in parentheses) are as follows:

- Percentage of pharmacist advice accepted by physicians (clinical pharmacy)
- Percentage of budget spent on anti-infectives (purchasing and inventory control including formulary)
- Inventory turnover rate (purchasing and inventory control including formulary)
- Size of formulary (purchasing and inventory control including formulary)

10.8.5 Department: Food Services Excluding Therapeutic Nutrition

Some of the quality or performance indicators (along with their corresponding principal functions in parentheses) are as follows:

- Number of food trays returned untouched (patient food service)
- Number of employee incidents (sanitation)
- Number of discrepancies of public health inspection (sanitation)
- Food waste, in dollars, per month (purchasing/inventory control)

10.9 PROBLEMS

1. Write a short essay on the history of quality in health care.
2. Discuss at least five reasons for escalating health care costs.
3. Discuss four important goals of quality in health care.
4. Describe at least seven quality dimensions of today's health care business.
5. Make a comparison of quality assurance and quality improvement in health care organizations.
6. Compare health care quality with industrial quality.
7. List five common appeals of health care administrators for total quality management and their corresponding physicians' reactions.
8. What is the Six Sigma methodology?
9. What are the important benefits of implementing the Six Sigma methodology in hospitals?
10. What are the potential barriers to the implementation of Six Sigma programs in health care organizations?

REFERENCES

1. Graham, N.O. Quality Trends in Healthcare. In *Quality in Healthcare: Theory, Application, and Evolution*, edited by N.O. Graham, 3–14. Gaithersburg, MD: Aspen Publishers, 1995.
2. Flexner, A. Medical Education in the United States and Canada. Report, Carnegie Foundation for the Advancement of Teaching. New York: Carnegie Foundation Bulletin, 1910.
3. Codman, E.A. The Product of the Hospital. *Surgical Gynecology and Obstetrics* 18 (1914): 491–496.
4. Gaucher, E.J., Coffey, R.J. *Total Quality in Health Care: From Theory to Practice.* San Francisco: Jossey-Bass Publishers, 1993.
5. Marszalek-Gaucher, E., Coffey, R.J. *Transforming Health Care Organizations: How to Achieve and Sustain Organizational Excellence.* New York: John Wiley & Sons, 1990.
6. Coile, R.C. *The New Hospital: Future Strategies for a Changing Industry.* Rockville, MD, Aspen Systems Publishers, 1986.
7. Schick, F.L., ed. *Statistical Handbook on Aging Americans.* Phoenix, AZ: Oryx Press, 1986.
8. Omachonu, V.K. *Total Quality and Productivity Management in Healthcare Organizations.* Milwaukee, WI: American Society for Quality Control, 1991.
9. Durbin, S., Haglund, C., Dowling, W. Integrating Strategic Planning and Quality Management in a Multi-Institutional System. In *Quality in Healthcare: Theory, Application, and Evolution*, edited by N.O. Graham, 231–248. Gaithersburg, MD: Aspen Publishers, 1995.
10. McLaughlin, C.P., Kaluzny, A.D. Defining Quality Improvement. In *Continuous Quality Improvement in Health Care*, edited by C.P. McLaughlin, A.D., Kaluzny, 3–40. Boston: Jones and Bartlett Publishers, 2006.

11. Cotlin, K.L., Aronow, D.B. Quality Assurance and Quality Improvement in the Information Age. In *Quality in Health Care: Theory, Application, and Evolution,* edited by N.O. Graham, 223–230. Gaithersburg, MD: Aspen Publishers, 1995.
12. Andrew, S.L. QA versus QI: The Changing Role of Quality in Health Care. *Journal of Quality Assurance* January/February (1991): 14, 15, 38.
13. Berwick, D.M. Peer Review and Quality Management: Are They Compatible? *Quality Review Bulletin* 16 (1990): 246–251.
14. Fainter, J. Quality Assurance Not Quality Improvement. *Journal of Quality Assurance* January/February (1991): 8, 9, 36.
15. Laffel, G., Blumenthal, D. The Case for Using Industrial Quality Management Science in Health Care Organization. *Journal of the American Medical Association* 262 (1989): 2869–2873.
16. Kratochwill, E.W., Sonda, L.P. Physician Involvement. In *Total Quality in Healthcare: From Theory to Practice*, edited by E.J. Gaucher, R.J. Coffey, 181–216. San Francisco: Jossey-Bass Publishers, 1993.
17. Stamatis, D.H. *Total Quality Management in Healthcare.* Chicago: Irwin Professional Publishing, 1996.
18. Frings, G.W., Graut, L. Who Moved My Sigma – Effective Implementation of the Six Sigma Methodology to Hospitals. *Quality and Reliability Engineering International* 21 (2005): 311–328.
19. Wilson, C.R.M. *Strategies in Healthcare Quality.* Toronto, Ontario: W.B. Saunders Company Canada Limited, 1992.

11 Quality Methods for Use in Health Care

11.1 INTRODUCTION

Over the years increasing attention is being given to improve quality of products and services in the industrial sector. This has resulted in the development of many quality improvement methods. Some examples of these methods are control chart, quality function deployment (QFD), and cause-and-effect diagram.[1,2] The application of methods such as these has been quite successful in improving the quality of industrial products.

As the result of their successful application in the industrial area, nowadays increasing attention is being paid to use these methods to improve quality in the area of health care. The quality methods can be classified under four distinct categories (examples of the methods belonging to each category are presented in parentheses): idea generation (e.g., brainstorming, multivoting, mind mapping, and morphological analysis), management (e.g., quality function deployment, Pareto analysis, quality circles, and benchmarking), analytical (e.g., cause-and-effect diagram, force field analysis, and fault tree analysis), and data collection, analysis, and display (e.g., check sheets, flowcharts, scatter diagrams, and statistical process control).

This chapter presents the quality methods belonging to each of the above four categories considered useful to improve quality in health care.[1-4]

11.2 GROUP BRAINSTORMING

This is probably the most widely used approach to generate new ideas in industry. A group of people participate in brainstorming sessions and in these sessions, one idea for finding solutions to a given problem triggers another idea and the process continues.[5] It is to be noted that the participants in the sessions belong to different backgrounds but have similar interests. Some useful guidelines to conduct effective brainstorming sessions are as follows[6-8]:

- Record ideas.
- Do not allow any criticism whatsoever during sessions.
- Aim to keep the ranks of participating individuals fairly equal.
- Welcome freewheeling and choose the timing of sessions with care.
- Combine and improve ideas after sessions.

The team concerned with improving health care quality can use this approach to get the ideas organized into a quality method such as a process flow diagram or a cause-and-effect diagram. Past experiences indicate that questions such as those listed below could be quite useful to start a health care quality-related session.[9]

- What are the main obstacles to improving quality?
- What are the organization's three most pressing unsolved quality-related problems?
- What type of action plan is necessary for overcoming these problems?
- What are the most pressing specific areas that need such an action plan?

11.3 CAUSE-AND-EFFECT DIAGRAM

This was developed by a Japanese quality expert named Kaoru Ishikawa.[10] Sometimes the cause-and-effect diagram is also called a fishbone diagram because of its resemblance to the skeleton of a fish, as shown in Figure 11.1. More specifically, the right-hand side of the diagram, i.e., the fish head, represents effect, and the left-hand side represents all the possible causes which are connected to the central line called the "Fish Spine."

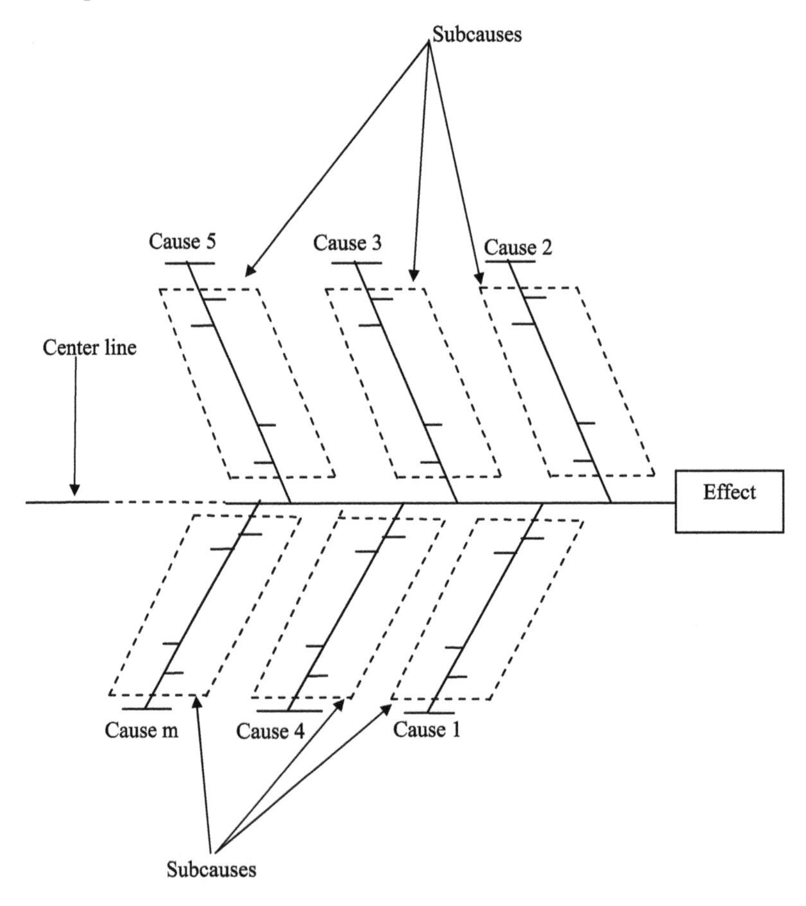

FIGURE 11.1 A cause-and-effect diagram.

The cause-and-effect diagram has proven to be an extremely useful tool in determining the root causes of a specified problem and to generate relevant ideas. With regard to, say, total quality management (TQM), customer satisfaction could be the effect and manpower, methods, machinery, and materials, the major causes. Usually, the following five steps are followed to develop a cause-and-effect diagram:

Step 1: Establish problem statement.
Step 2: Brainstorm to identify all possible causes.
Step 3: Developmajor cause groups by stratifying into natural groupings and process steps.
Step 4: Constructthe diagram by linking the identified causes under appropriate process steps and fill in the effect or the problem in box (i.e., the fish head) on the extreme right-hand side of the diagram.
Step 5: Refinecause groups or categories by asking questions such as "what causes this?" and "why does this condition exist?"

Additional information on the cause-and-effect diagram is available in Chapter 4.

11.4 QUALITY FUNCTION DEPLOYMENT

This method was developed in Japan and is used to optimize the process of developing and manufacturing new products per customer requirements.[11,12] It can also be used to improve quality in the area of health care. Thus, in a broader context, QFD may simply be expressed as a formal process used to translate the needs of customers into a set of technical requirements.

QFD makes use of a set of matrices for relating customer needs to counterpart characteristics expressed as technical specifications and process control requirements. The important QFD planning documents are as follows[13]:

- **Customer requirements planning matrix.** This translates the consumer requirements into product counterpart characteristics.
- **Product characteristic deployment matrix.** This translates final product counterpart characteristics into critical component characteristics.
- **Operating instructions.** They identify operations that must be accomplished successfully to achieve critical parameters.
- **Process plan and quality control charts.** They identify important process and product parameters along with control limits.

It is to be noted that often a QFD matrix is called the "House of Quality" because of its resemblance to the structure of a house. The following six steps are required to build the house of quality[11,12]:

Step 1: Identify customer needs.
Step 2: Identifythe essential process/product characteristics that will satisfy the customer needs.

Step 3: Establishappropriate relationships between the customer requirements and the counterpart characteristics.
Step 4: Perform analysis of competing products.
Step 5: Establishcounterpart characteristics of competing products and formulate goals.
Step 6: Highlightcounterpart characteristics to be utilized in the remaining process.

One important benefit of QFD is that it helps to encourage organizations to focus on the process itself rather than focusing on the product or service.

11.5 PROCESS FLOW DIAGRAM

This is one of the most widely used methods in the quality improvement process. A process flow diagram illustrates and clarifies tasks/events associated with a process as well as the tasks/events between them.[3,9] Thus, it is a useful tool to highlight the existing situation, differences between what should or is thought to be happening and the actual situation or condition, potential problem areas, and the proposed situation or condition.

The following steps are followed in constructing a process flow diagram[9]:

Step 1: Form a team of appropriate individuals.
Step 2: Define the process and its associated boundaries.
Step 3: Brainstorm the process under consideration.
Step 4: Use the simplest symbols possible in the diagram construction.
Step 5: Draw the steps followed by the process.
Step 6: Ensure that each feedback loop is accounted for.
Step 7: Draw all the appropriate steps the process under consideration should follow.
Step 8: Review and make adjustments as considered appropriate.

Additional information on this method is available in refs. [3, 9].

11.6 AFFINITY DIAGRAM

The affinity diagram was developed by a Japanese anthropologist named Jiro Kawakita in the 1960s. It may be described as a group decision-making approach designed to sort a large number of possible ideas, opinions, concepts, and process variables into naturally compatible groups.

In order to develop an affinity diagram one has to perform steps such as those listed below.

1. Sort the brainstormed list of ideas.
2. Move ideas from the brainstormed list into affinity sets.
3. Create groups of related ideas.

Some useful guidelines associated with this method are as follows[3]:

- Ensure that ideas are described with phrases or sentences.
- Aim to establish around 5 to 10 groups.
- If any group is much bigger than other groups, consider splitting it.

All in all, affinity diagram can be a useful tool to improve quality in the health care area. Additional information on this method is available in ref. [14].

11.7 CHECK SHEET

This is a useful generic tool that can be adapted to a wide range of purposes, including improving quality in health care. A check sheet may be described as a structured, prepared form for collecting and analyzing data. In quality work, check sheets are normally used during a quality improvement process to collect frequency-related data later displayed in a Pareto diagram.[3] More specifically, they are used in situations such as listed below.

- When it is possible to observe and collect data repeatedly by the same individual or at the same location
- When gathering data from a production process facility
- When collecting data on the frequency or patterns of events, defect causes, problems, and so on

Four basic steps for constructing a check sheet are as follows[4]:

Step 1: Agree on the item to be observed.
Step 2: Decide on time interval for collecting data.
Step 3: Design a simple and straightforward form.
Step 4: Collect the appropriate data consistently.

Additional information on this approach is available in refs. [10, 15].

11.8 FORCE FIELD ANALYSIS

This method was developed by social psychologist Kurt Lewin (1890–1947) to identify forces that are concerned with a specific issue under consideration.[16,17] Sometimes the method is also called barriers and aids analysis.[3] In any case, the method calls for writing the problem/issue statement at the top of a sheet of paper and creating two columns below it for writing negative forces on one side and the positives on the other. Subsequently, the forces are ranked and ways for mitigating the negative forces and accentuating the positive forces are explored.

More specifically, the six steps shown in Figure 11.2 are followed in performing force field analysis.[3]

Additional information on the force field analysis is available in refs. [3, 8, 16].

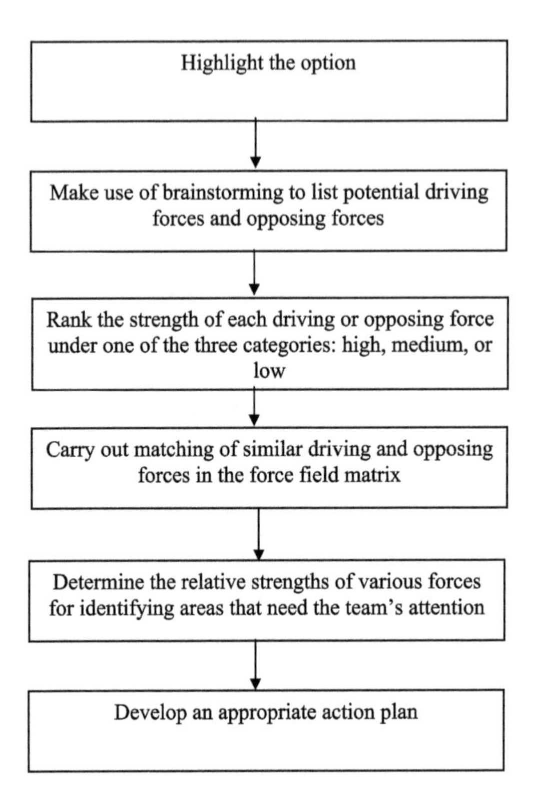

FIGURE 11.2 Steps for performing force field analysis.

11.9 MULTIVOTING

This is an effective approach used to reduce a large number of ideas to a manageable few considered important by the participating individuals. Usually when using this method, the number of ideas is reduced to three to five.[3] Another thing that can be added about this method is that it is a form of convergent thinking because the main objective is to cut down the number of ideas being considered.

The seven main steps involved in conducting the multivoting activity are as follows:

Step 1: Conduct brainstorming to generate a list of ideas or options.
Step 2: Review the list generated by the brainstorming activity.
Step 3: Make participants vote for the ideas considered to be worthy of further discussion.
Step 4: Highlight items for next round of voting.
Step 5: Vote again.
Step 6: Repeat steps 4 and 5 until options or ideas are reduced to three to five.
Step 7: Discuss remaining ideas or options and take appropriate actions.

Needless to say, multivoting can be an effective tool for improving quality in health care. Additional information on this approach is available in ref. [17].

11.10 PARETO CHART

The Pareto chart belongs to one of the seven basic tools of quality control: control chart, cause-and-effect diagram, flowchart, scatter diagram, histogram, check sheet, and Pareto chart. It is named after Italian economist and sociologist Vilfredo Pareto (1848–1923), who conducted a study concerning the spread of wealth and poverty in Europe in the early 1900s. He concluded that wealth was concentrated in the hands of approximately 20% of the people and poverty in approximately 80%. The findings of Pareto may be called the law of the "significant few versus the trivial many."

The use of a Pareto chart or principle in quality assurance was popularized by Joseph M. Juran, who believed that around 80% of the quality-related scrap was caused by about 20% of the problems. A Pareto chart may be described as a special kind of bar chart where the plotted values are arranged in descending order.

The following steps are associated with the construction of a Pareto chart[13,18]:

1. Determine the method of classifying the data (e.g., by problem, type of nonconformity, etc.).
2. Decide what is to be used in ranking the characteristics (e.g., frequency or dollars).
3. Collect all necessary data (i.e., for a time interval).
4. Summarize the data values.
5. Rank all classifications from largest to smallest.
6. Calculate the cumulative percentage, if required.
7. Construct the Pareto chart.
8. Determine the vital or significant few.

11.11 SCATTER DIAGRAM

This diagram is used to analyze relationships between two variables. In this diagram, one variable is plotted on the horizontal axis and the other on the vertical axis. Their intersecting points' pattern can graphically show relationship patterns. A scatter diagram is frequently used in proving or disproving cause-and-effect relationships. It is to be noted, although this diagram depicts relationships, it cannot by itself prove that one variable causes the other.

In quality control work, a scatter diagram is a quite useful tool for designing a control system to ensure that gains from quality improvement efforts are maintained. A scatter diagram can be constructed by following the three steps listed below.[9]

1. Gather minimum of 50 paired samples of data.
2. Draw horizontal and vertical axes. Ensure that the values on these axes become higher as one moves away from the origin and aim to make the horizontal axis as the expected cause and the vertical axis as the effect.
3. Plot all the data values.

All in all, the scatter diagram can be a useful tool to improve quality in the area of health care. Additional information on this method is available in refs. [9, 17].

11.12 CONTROL CHART

This chart is widely used in quality control work and it may be described as a graphical method used to determine whether a process is in a "state of statistical control" or out of control.[19] Sometimes, the control chart is also called the Shewhart chart or process-behavior chart. Figure 11.3 shows the basic form of a control chart. As shown in the figure, it is basically composed of three items: upper control limit (UCL), lower control limit (LCL), and mean or standard value of the characteristic of interest.

Typically, random samples of values taken from an ongoing process are plotted on the control chart shown in Figure 11.3. When any of these values falls outside the upper and lower control limits, it simply indicates that the ongoing process is out of statistical control and some corrective action is necessary.

Although control charts were originally developed for use in manufacturing areas, they can easily be tailored for applications in service organizations; for example, in organizations (along with respective typical quality measures in parentheses) such as a hospital (on-time delivery of medication and meals, laboratory test accuracy, etc.), ambulance (response time), and post office (percentage express mail delivered on time, sorting accuracy).[19,20]

There are many different types of control charts including c-charts, p-chart, $\bar{x}$- chart, R-chart, and EWMA chart (exponentially weighted moving average chart).[13,21] The c-chart is described below.

11.12.1 The C-Chart

This chart is used to control the occurrence of the number of defects per unit when the subgroup size does not vary or remains constant and the chances for defect occurrence in an item are high. The c-chart mathematical calculations are based on

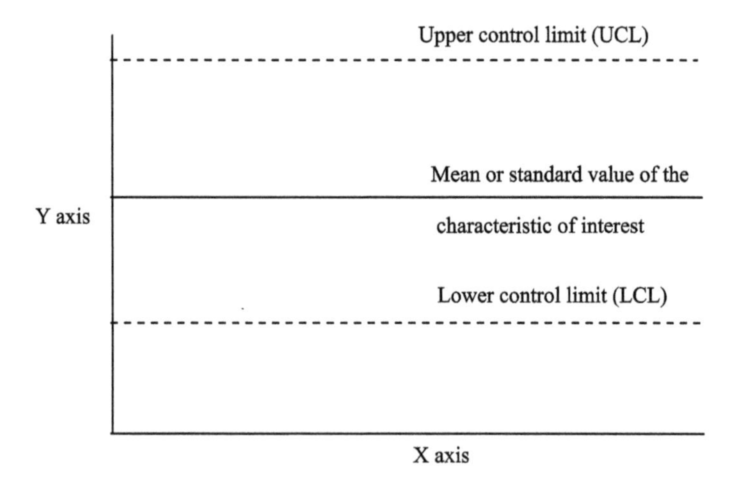

FIGURE 11.3 A sketch of a basic control chart.

the Poisson distribution. Thus, the mean, m, of the Poisson distribution in the term of c-chart is expressed by[13]:

$$m = \frac{\theta}{\alpha} \tag{11.1}$$

where

θ is the total number of defects.
α is the total number of items/units.

Similarly, the standard deviation is given by:

$$\sigma = (m)^{1/2} \tag{11.2}$$

where

σ is the standard deviation of the Poisson distribution.

Thus, the upper and lower control limits of the c-chart are given by[20]:

$$UCL = m + 3\sigma \tag{11.3}$$

and

$$LCL = m - 3\sigma \tag{11.4}$$

where

UCL is the upper control limit of the c-chart
LCL is the lower control limit of the c-chart.

Example 11.1

Assume that in a hospital during a 12-month period a total of 191 health-related accidents/incidents occurred; their monthly breakdowns are presented in Table 11.1. Develop the c-chart.

By substituting the above specified values into Equation (11.1) we get:

$$m = \frac{191}{12} = 15.92 \textit{ accidents / incidents per month}$$

Using the above calculated value in Equation (11.2) yields:

$$\sigma = (15.92)^{1/2} = 3.99$$

Table 11.1 Hospital Monthly Accident/Incident Occurrences

Month	No. of Accidents/Incidents
January (J)	15
February (F)	15
March (M)	20
April (A)	25
May (MY)	10
June (JN)	18
July (JY)	21
August (AU)	15
September (S)	20
October (O)	10
November (N)	14
December (D)	8

Using the above two calculated values in Equations (11.3) and (11.4), we get:

$$UCL = 15.92 + 3(3.99) = 27.89$$

and

$$LCL = 15.92 - 3(3.99) = 3.95$$

Using the above results and the specified data values in Table 11.1, the Figure 11.4 c-chart was developed. The chart shows that all specified data values in Table 11.1 are within the upper and lower control limits. It simply means that there is no abnormality.

11.13 PROBLEMS

1. Write an essay on quality control methods.
2. What is a control chart?
3. What are the seven basic tools of quality control?
4. Describe the group brainstorming method.
5. List and discuss steps involved in constructing a cause-and-effect diagram.
6. What are the other names used for the following three methods:
   - Control chart
   - Cause-and-effect diagram
   - Force field analysis
7. List at least five types of control charts.
8. Discuss quality function deployment.

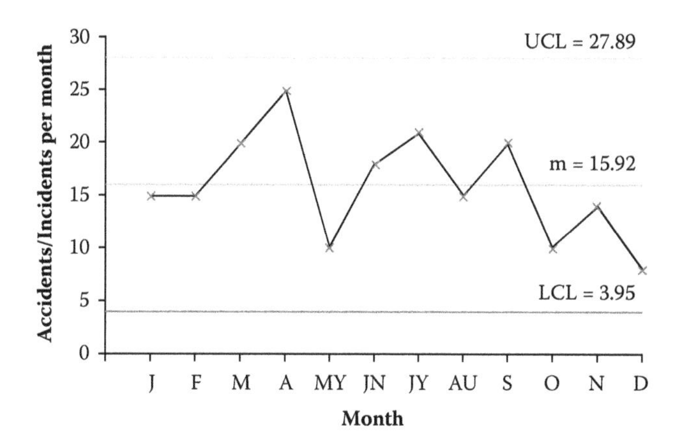

FIGURE 11.4 Chart for Example 11.1

9. Compare the Pareto chart with the scatter diagram.
10. Assume that in a hospital health-related accident/incident occurrence data over the past 12-month period were examined. During January, February, March, April, May, June, July, August, September, October, November, and December 20, 25, 10, 15, 18, 27, 9, 14, 16, 15, 17, and 30 health-related accidents/incidents occurred, respectively. Develop the c-chart.

REFERENCES

1. Mears, P. *Quality Improvement Tools and Techniques*. New York: McGraw-Hill, 1995.
2. Kanji, G.K., Asher, M. *100 Methods of Total Quality Management*. London: Sage Publications Ltd., 1996.
3. Gaucher, E.J., Coffey, R.J. *Total Quality in Health Care: From Theory to Practice*. San Francisco: Jossey-Bass Publishers, 1993.
4. Stamatis, D.H. *Total Quality Management in Health Care*. Chicago: Irwin Professional Publishing, 1996.
5. Osborn, A.F. *Applied Imagination*. New York: Charles Scribner's Sons, 1963.
6. Studt, A.C. How to Set Up Brainstorming Sessions. In *Management Guide for Engineers and Technical Administrators*, edited by N.P. Chironis, 276–277. New York: McGraw-Hill, 1969.
7. Dhillon, B.S. *Engineering and Technology Management Tools and Applications*. Boston: Artech House, 2002.
8. Dhillon, B.S. *Creativity for Engineers*. Hackensack, NJ: World Scientific Publishing, 2006.
9. Stamatis, D.H. *Total Quality Management in Healthcare*. Chicago: Irwin Professional Publishing, 1996
10. Ishikawa, K. *Guide to Quality Control*. Tokyo: Asian Productivity Organization, 1976.
11. Mizuno, S., Akao, Y., eds. *QFD: The Customer-Driven Approach to Quality Planning and Deployment*. Tokyo: Asian Productivity Organization, 1994.

12. Bossart, J.L. *Quality Function Deployment: A Practioner's Approach.* Milwaukee, WI: ASQC Quality Press, 1991.
13. Dhillon, B.S. *Reliability, Quality and Safety for Engineers.* Boca Raton, FL: CRC Press, 2005.
14. Pries, K.H. *Six Sigma for the Next Millennium, A CSSBB Guidebook.* Milwaukee, WI: American Society for Quality, 2006.
15. Pyzdek, T. *The Six Sigma Handbook: A Complete Guide for Green Belts, Black Belts, and Managers at All Levels.* New York: McGraw-Hill, 2003.
16. Jay, R. *The Ultimate Book of Business Creativity: 50 Great Thinking Tools, for Transforming Your Business.* Oxford, UK: Capstone Publishing Limited, 2000.
17. Tague, N.R. *The Quality Toolbox.* Milwaukee, WI: ASQ Press, 2005.
18. Besterfield, D.H. *Quality Control.* Upper Saddle River, NJ: Prentice Hall, 2001.
19. Rosander, A.C. *Applications of Quality Control in the Service Industries.* New York: Marcel Dekker, 1985.
20. Evans, J.R., Lindsay, W.M. *The Management and Control of Quality.* New York: West Publishing, 1996.
21. Ryan, T.P. *Statistical Methods for Quality Improvements.* New York: John Wiley & Sons, 2000.

Bibliography

Literature on Reliability Technology, Human Error, and Quality in Health Care

INTRODUCTION

Over the years, a large number of publications relating to reliability technology, human error, and quality in health care have appeared in the form of journal articles, conference proceedings articles, books, etc. This appendix presents an extensive list of such publications.

The period covered by the listing is from 1962 to 2006. The main objective of this listing is to provide readers with sources for obtaining additional information on reliability technology, human error, and quality in health care.

PUBLICATIONS

Adams, J.G., Bohan, S. System Contributions to Error. *Academic Emergency Medicine* 7 (2000): 1189–1193.

Al-Abdulla, H.M., Lulu, D.J. Hypokalaemia and Pacemaker Failure. *American Surgery* 40 (1974): 234–236.

Al-Assaf, A.F., Schmele, J.A. *The Textbook of Total Quality in Healthcare.* Delray Beach, FL: St. Lucie Press, 1993.

Alberti, K.G.M.M. Medical Errors: A Common Problem. *British Medical Journal* 322 (2001): 501–502.

Alexander, K., Clarkson, P.J. Good Design Practice for Medical Devices and Equipment, Part II: Design for Validation. *Journal of Medical Engineering and Technology* 24 (2000): 53–62.

Allen, D. California Home to Almost One-Fifth of U.S. Device Industry. *Medical Device & Diagnostic Industry*, 19 (1997): 64–67.

Allen, R.C. FDA and the Cost of Health Care. *Medical Device & Diagnostic Industry* 18 (1996): 28–35.

Altshuler, C., Aster, R., Covino, K., Gordon, S., Gajewski, M., Smith, A., Horrigan, S., Pfaff, K. Automated Donor-Recipient Identification Systems as a Means of Reducing Human Error in Blood Transfusion. *Transfusion* 17 (1977): 586–597.

Andrews, L.B., Stocking, C., Krizek, T., Gottlieb, L., Krizek, C., Vargish, T., Siegler, M. An Alternative Strategy for Studying Adverse Events in Medical Care. *Lancet* 349 (1997): 309–313.

Angelo, M.D. Internet Solution Reduces Medical Errors. *Health Management Technology,* February (2000): 20–21.

Anon. The Quality of the NHS. *Quality World* 29 (2003): 14.

Appler, W.D., McMann, G.L. Medical Device Regulation: The Big Picture. In *The Medical Device Industry: Science, Technology, and Regulation in a Competitive Environment,* edited by N.F. Estrin, 35–51. New York: Marcel Dekker, 1990.

Arcarese, J. An FDA Perspective. Proceedings of the First Symposium on Human Factors in Medical Devices, 1989, 23–24.

Arearese, J.S. FDA's Role in Medical Device User Education. *The Medical Device Industry: Science, Technology, and Regulation in a Competitive Environment*, edited by N.F. Estrin, 129–138. New York: Marcel Dekker, 1990.

Arnstein, F. Catalogue of Human Error. *British Journal of Anaesthesia* 79 (1997): 645–656.

Association for the Advancement of Medical Instrumentation. *Human Factors Engineering Guidelines and Preferred Practices for the Design of Medical Devices*. ANSI/AAMI HE-48. Arlington, VA: Association for the Advancement of Medical Instrumentation, 1993.

Banta, H.D. The Regulation of Medical Devices. *Preventive Medicine* 19 (1990): 693–699.

Barker, K.N., McConnell, W.E. Detecting Errors in Hospitals. *American Journal of Hospitals* 19 (1962): 361–369.

Bassen, H., Silberberg, J., Houston, F., Knight, W. Computerized Medical Devices, Trends, Problems, and Safety. *IEEE Aerospace and Electronic Systems (AES)* September (1986): 20–24.

Batalden, P. Deming Offers Much to Healthcare Workers. *Quality Progress* 35 (2002): 10–12.

Bates, D.W., Cullen, D.J., Small, S.D., Cooper, J.B., Nemeskal, A.R., Leape, L.L. The Incident Reporting System Does Not Detect Adverse Drug Events: A Problem for Quality Improvement. *Joint Commission Journal on Quality Improvement* 21 (1995): 541–548.

Bates, D.W., Spell, N., Cullen, D.J. The Costs of Adverse Drug Events in Hospitalized Patients. *Journal of the American Medical Association* 277 (1997): 307-311.

Bates, D.W. Using Information Technology to Reduce Rates of Medication Errors in Hospitals. *British Medical Journal* 320 (2000): 788–791.

Battles, J., Mercer, Q., Whiteside, M., Bradley, J. A Medical Event Reporting System for Human Errors in Transfusion Medicine. In *Advances in Applied Ergonomics*, edited by A.F. Ozok, G. Salvendy, 809–814. West Lafayette, IN: USA Publishing, 1996.

Battles, J.B., Kaplan, H.S., Van der Schaaf, T.W., Shea, C.E. The Attributes of Medical Event-Reporting Systems. *Archives of Pathology and Laboratory Medicine* 122 (1998): 231–238.

Baylis, F. Errors in Medicine: Nurturing Truthfulness. *Journal of Clinical Ethics* 8 (1997): 336–340.

Beasley, L.J. Reliability and Medical Device Manufacturing. Proceedings of the Annual Reliability and Maintainability Symposium, 1995, 128–131.

Beckmann, U., Baldwin, I., Hart, G.K., Runciman, W.B. The Australian Incident Monitoring Study in Intensive Care: AIMS-ICU. An Analysis of the First Year of Reporting. *Anaesthesia and Intensive Care* 24 (1996): 320–329.

Beier, B. Liability and Responsibility for Clinical Medical Software in the Federal Republic of Germany. Proceedings of the 10th Annual Symposium on Computer Applications Medical Care, 1986, 364–368.

Beiser, E.N. Reporting Physicians' Mistakes. *Rhode Island Medical Journal* 73 (1990): 333–336.

Beith, B.H. Human Factors and the Future of Telemedicine. *Medical Device & Diagnostic Industry* 21 (1999): 36–40.

Belkin, L. Human and Mechanical Failures Plague Medical Care. *New York Times,* March 31, 1992, B1, B6.

Bell, B.M. Error in Medicine. *Journal of the American Medical Association* 274 (1995): 457–461.

Bell, D.D. Contrasting the Medical-Device and Aerospace-Industries Approach to Reliability. Proceedings of the Annual Reliability and Maintainability Symposium, 1996, 125–127.

Bell, R., Krivich, M.J. *How to Use Patient Satisfaction Data to Improve Healthcare Quality.* Milwaukee, WI: ASQ Quality Press, 2000.

Ben-Zvi, S. Quality Assurance in Transition. *Biomedical Instrumentation & Technology* 23 (1989): 27–33.

Berglund, S. Systems Failures, Human Error and Health Care. *Medical Liability Monitor* (1998): 1–4.

Berndt, D.J., Fisher, J.W., Hevner, A.R. Healthcare Data Warehousing and Quality Assurance. *Computer* 34 (2001): 56–65.

Bethune, J. The Cost-Effective Bugaboo. *Medical Device & Diagnostic Industry* 19 (1997): 12–15.

Beuscart, R.J., Alao, O.O., Brunetaud, J.M. Health Telematics: A Challenge for Healthcare Quality. Proceedings of the 23rd Annual International Conference of the IEEE Engineering in Medicine and Biology Society, 2001, 4105–4107.

Beyea, S.C. Tracking Medical Devices to Ensure Patient Safety. *AORN Journal* 77 (2003): 192–194.

Billings, C.E., Woods, D.D. Human Error in Perspective: The Patient Safety Movement. *Postgraduate Medicine* 109 (2001): 23–25.

Bindler, R., Boyne, T. Medication Calculation Ability of Registered Nurses. *Image* 23 (1991): 221–224.

Blum, L.L. Equipment Design and "Human" Limitations. *Anesthesiology* 35 (1971): 101–102.

Blumenthal, D. Making Medical Errors into Medical Treasures. *Journal of the American Medical Association* 272 (1994): 1867–1869.

Bogner, M.S. Designing Medical Devices to Reduce the Likelihood of Error. *Biomedical Instrumentation & Technology* 33 (1999): 108–113.

Bogner, M.S., ed. *Human Error in Medicine.* Hillsdale, NJ: Lawrence Erlbaum Associates, 1994.

Bogner, M.S. Helping to Reduce Human Error in Health Care Technology. *Biomedical Instrumentation and Technology* 47 (2003): 61–64.

Bogner, M.S. Human Error in Medicine: A Frontier for Change. In *Human Error in Medicine*, edited by M.S. Bogner, 373–383. Hillsdale, NJ: Lawrence Erlbaum Associates, 1994.

Bogner, M.S. Human Factors, Human Error and Patient Safety Panel. Proceedings of the Annual Human Factors Society Conference 2, 1998, 1053–1057.

Bogner M.S. Medical Device and Human Error, Human Performance. In *Automated Systems: Current Research and Trends*, edited by Mouloua, M., Parasuraman, P., 64–67. Hillsdale, NJ: Lawrence Erlbaum Associates, 1994.

Bogner, M.S. Medical Device: A Frontier for Human Factors. *CSERIAC Gateway* IV (1993): 12–14.

Bombino, A. Lessons Learned in Applying QFD at Baxter Healthcare. Proceedings of the NEPCON West '92 Conference, 1992, 877–880.

Boumil, M.M., Elias, C.E. *The Law of Medical Liability in a Nutshell.* St. Paul, MN: West Publishing, 1995.

Bousvaros, G.A., Don, C., Hopps, J.A. An Electrical Hazard of Selective Angiocardiography. *Canadian Medical Association Journal* 87 (1962): 286–288.

Bowling, P., Berglund, R. HIPAA: Where Healthcare and Software Quality Meet. Proceedings of the 57th Annual Quality Congress: Expanding Horizons, 2003, 271–281.

Bracco, D., Favre, J.B., Bissonnette, B. Human Errors in a Multidisciplinary Intensive Care Unit: A 1-Year Prospective Study. *Intensive Care Medicine* 27 (2001): 137–145.

Bradley, J.B. Controlling Risk in a Complex Environment: The Design of Computer Based Systems to Minimize Human Error in Medicine. Proceeding of the ASIS Annual Conference, October 1996, 210–213.

Brady, W.J., Perron, A. Errors in Emergency Physician Interpretation of ST-Segment Elevation in Emergency Department Chest Pain Patients. *Academic Emergency Medicine* 7 (2000): 1256–1260.

Brasel, K.J., Layde, P.M., Hargarten, S. Evaluation of Error in Medicine: Application of a Public Health Model. *Academic Emergency Medicine* 7 (2000): 1298–1302.

Bratman, R.L. A National Database of Medical Error. *Journal of Royal Society of Medicine* 93 (2000): 106.

Brazeau, C. Disclosing the Truth About a Medical Error. *American Family Physician* 60 (1999): 1013–1014.

Brennan, T.A. The Institute of Medical Report on Medical Errors—Could It Do Harm? *New England Journal of Medicine* 342 (2000): 1123–1125.

Brennan, T.A., Bates, D.W., Pappius, E., Kuperman, G.J., Sittig, D., Burstin, H., et al. Using Information Systems to Measure and Improve Quality. *International Journal of Medical Informatics* 53 (1999): 115–124.

Brennan, T.A., Leape, L.L., Laird, N.M. Incidence of Adverse Events and Negligence in Hospitalised Patients: Results of the Harvard Medical Practice Study I. *New England Journal of Medicine* 324 (1991): 370-376.

Brennan, T.A., Leape, L.L., Laird, N.M. Incidence of Adverse Events and Negligence in Hospitalised Patients: Results of the Harvard Medical Practice Study II. *New England Journal of Medicine* 324 (1991): 377-384.

Brennan, T.A., Localio, A.R., Laird, N.M. Reliability and Validity of Judgments Concerning Adverse Events and Negligence. *Medical Care* 27 (1989): 1148–1158.

Brennan, T.A., Localio, R., Leape, L.L., Laird, N.M. Identification of Adverse Events Occurring During Hospitalisation. *Annals of Internal Medicine* 112 (1990): 221–226.

Brienza, J. Medical Mistakes Study Highlights Need for System Wide Improvements. *Trail* February (2000): 15–17.

Brooke, P.S. Shaping the Medical Error Movement. *Nursing Management* 31 (2000): 18–19.

Brown, P.J.B., Warmington, V. Data Quality Probes—Exploiting and Improving the Quality of Electronic Patient Record Data and Patient Care. *International Journal of Medical Informatics* 68 (2002): 91–98.

Brown, R.W. Errors in Medicine. *Journal of Quality Clinical Practice* 17 (1997): 21–25.

Bruley, M.E. Ergonomics and Error: Who Is Responsible. Proceeding of the First Symposium on Human Factors in Medical Devices, 1989, 6–10.

Bruner, J.M.R. Hazards of Electrical Apparatus. *Anesthesiology* 28 (1967): 396–425.

Burchell, H.B. Electrocution Hazards in the Hospital or Laboratory. *Circulation* (1963): 1015–1017.

Burlington, D.B. Human Factors and the FDA's Goal: Improved Medical Device Design. *Biomedical Instrumentation & Technology* 30 (1996): 107–109.

Busse, D.K., Johnson, C.W. Human Error in an Intensive Care Unit B: A Cognitive Analysis of Critical Incidents. Proceedings of the 17th International Conference of the Systems Safety Society, Orlando, Florida, August 1999, 70–75.

Busse, D.K., Wright, D.J. Classification and Analysis of Incidents in Complex, Medical Environment. *Topics in Health Information Management* 20 (2000): 56–58.

Caldwell, R. Alerting Staff to Medication Errors. *Health Management Technology* August (2000): 52.

Camishion, R.C. Electrical Hazards in the Research Laboratory. *Journal of Surgical Research* 6 (1966): 221–227.

Caplan, R.A., Vistica, M.F., Posner, K.L., Cheney, F.W. Adverse Anesthetic Outcomes Arising from Gas Delivery Equipment: A Closed Claims Analysis. *Anesthesiology* 87 (1997): 741–748.

Caroselli, M., Edison, L. *Quality Care: Prescriptions for Injecting Quality into Healthcare Systems*. Boca Raton, FL: St. Lucie Press, 1997.

Carstens, D., Hollingsworth, A.T., Barlow, J., Bean, L.A., Hott, D., Pierce, B. The Technology Gap in Patient Safety. Proceedings of the 8th World Multi-Conference on Systemics, Cybernetics and Informatics 4, 2004, 268–272.

Casarett, D., Helms, C. Systems Errors versus Physicians' Errors: Finding the Balance in Medical Education. *Academic Medicine* 74 (1999): 19–22.

Casey, S. *Set Phasers on Stun: And Other True Tales of Technology and Human Error.* Santa Barbara, CA: Argean, 1993.

Cattaneo, C.R., Vecchio, A.D. Human Errors in the Calculation of Monitor Units in Clinical Radiotherapy Practice. *Radiotherapy & Oncology* 28 (1993): 86–88.

Chae, Y.M., Kim, H.S., Tark, K.C. Analysis of Healthcare Quality Indicator Using Data Mining and Decision Support System. *Expert Systems with Applications* 24 (2003): 167–172.

Chaplin, E. Customer Driven Healthcare Comprehensive Quality Function Deployment. Proceedings of the 56th Annual Quality Congress, 2002, 767–781.

Chassin, M.R. Is Health Care Ready for Six Sigma Quality. *Milbank Quarterly: Journal of Public Health and Health Care Policy* 76 (1998): 59–61.

Chassin, M.R., Galvin, R.W. The Urgent Need to Improve Health Care Quality. *Journal of the American Medical Association* 280 (1998): 1000–1005.

Chen, Y., Agarwal, B.D., So, P.K. The Role of Numerical Stress Analysis in Failure Analysis of a Medical Device. *Plastics Engineering* 55 (1999): 45–47.

Chopra, V., Bovill, J.G., Koornneef, J.F. Reported Significant Observations During Anaesthesia: A Prospective Analysis Over an 18-month Period. *British Journal of Anaesthesia* 68 (1992): 13–17.

Christensen, J.F., Levinson, W., Dunn, P.M. The Heart of Darkness: The Impact of Perceived Mistakes on Physicians. *Journal of General Internal Medicine* 7 (1992): 424–431.

Chung, P.H., Zhang, J., Jonson, T.R., Patel, V.L. An Extended Hierarchical Task Analysis for Error Prediction in Medical Devices. Proceedings of the AMIA Annual Symposium, 2003, 165–169.

Classen, D. Patient Safety, the Name is Quality. *Trustee* 53 (2000): 12–15.

Clayton, M. The Right Way to Prevent Medication Errors. *Registered Nurse (RN)* June (1987): 30–31.

Clifton, B.S., Hotten, W.I.T. Deaths Associated with Anaesthesia. *British Journal of Anaesthesia* 35 (1963): 250–259.

Cohen, M.R., Senders, J, Davis, N.M. Failure Mode and Effects Analysis: A Novel Approach to Avoiding Dangerous Medication Errors and Accidents. *Hospital Pharmacy* 29 (1994): 319–330.

Cohen, T. Computerized Maintenance Management Systems: How to Match Your Department's Needs with Commercially Available Products. *Journal of Clinical Engineering* November/December (1995): 457–461.

Cohen, T. Validating Medical Equipment Repair and Maintenance Metrics: A Progress Report. *Biomedical Instrumentation & Technology* January/February (1977): 23–32.

Cole, T. Medical Errors vs Medical Injuries: Physicians Seek to Prevent Both. *Journal of the American Medical Association* 284 (2000): 2175–2176.

Coleman, I.C. Medication Errors: Picking up the Pieces. *Drug Topics* March 15, 1999, 83–92.

Conlan, M.F. IOM's Med-Error Report Spurs Changes Among Pharmacies. *Drug Topics* 144 (2000): 70–71.

Cook, R.I., Woods, D.D. Operating at the Sharp End: The Complexity of Human Error. In *Human Error in Medicine*, edited by M.S. Bogner, 255–310. Hillsdale, NJ: Lawrence Erlbaum Associates, 1994.

Cooper, D.M. Series of Errors Led to 300 Unnecessary Mastectomies. *British Medical Journal* 320 (2000): 597.

Cooper, J.B. Toward Prevention of Anesthetic Mishaps. *International Anesthesiology Clinics* 22 (1985): 167–183.

Cooper, J.B., Gaba, D.M. A Strategy for Preventing Anesthesia Accidents. *International Anesthesiology Clinics* 27 (1989): 148–152.

Cooper, J.B., Newbower, R.S., Kitz, R.J. An Analysis of Major Errors and Equipment Failures in Anesthesia Management: Consideration for Prevention and Detection. *Anesthesiology* 60 (1984): 34–42.

Cooper, J.B., Newbower, R.S., Long, C.D., McPeek, B. Preventable Anesthetic Mishaps: A Study of Human Factors. *Anesthesiology* 49 (1978): 399–406.

Cooper, J.B., Newbower, R.S., Ritz, R.J. An Analysis of Major Errors and Equipment Failures in Anesthesia Management: Considerations for Prevention and Detection. *Anesthesiology* 60 (1984): 34–42.

Cott, H.P. Human Error in Health Care Delivery: Cases, Causes, and Correction. Proceedings of the Human Factors and Ergonomics Society, 1993, 846–848.

Cott V.H. Human Error: Their Causes and Reduction. In *Human Error in Medicine*, edited by M.S. Bogner, 53–65. Hillsdale, NJ: Lawrence Erlbaum Associates, 1994.

Craig, J., Wilson, M.E. A Survey of Anaesthetic Misadventures. *Anaesthesia* 36 (1981): 933–936.

Crane, M. How Good Doctors Can Avoid Bad Errors. *Medical Economics,* April 28, 1997, 36–43.

Crowley, J.J. Identifying and Understanding Medical Device Use Errors. *Journal of Clinical Engineering* 47 (2002): 188–193.

Cullen, D.J. Risk Modification in the Post Anesthesia Care Unit. *International Anesthesiology Clinics* 27 (1989): 184–187.

Cuthrell, P. Managing Equipment Failures: Nursing Practice Requirements for Meeting the Challenges of the Safe Medical Devices Act. *Journal of Intravenous Nursing* 19 (1996): 264-268.

Dain, S. Normal Accidents: Human Error and Medical Equipment Design. *Heart Surgery Forum* 5 (2002): 254–257.

David, P.P., Glyyn, C.N., Laura, M. Increase in US Medication-Error Deaths Between 1983–1993. *Lancet* 351 (1998): 643–644.

Davis, N., Cohen, M. *Medication Errors: Causes and Prevention.* Huntingdon Valley, PA: N.M. Davis, 1983.

Dawson, N.V. Systematic Errors in Medical Decision Making: Judgment Limitations. *Journal of General Internal Medicine* 2 (1987): 183–187.

De Lemos, Z. FMEA Software Program for Managing Preventive Maintenance of Medical Equipment. Proceedings of the IEEE 30th Annual Northeast Bioengineering Conference, 2004, 247–248.

De Leval, M.R. Human Factors and Surgical Outcomes: A Cartesian Dream. *Lancet* 349 (1997): 723–726.

De Leval, M.R., Reason, J.T. The Human Factor in Cardiac Surgery: Errors and Near Misses in a High Technology Medical Domain. *Annals of Thoracic Surgery* 72 (2001): 300–305.

DeAnda, A., Gaba, D.M. Unplanned Incidents During Comprehensive Anesthesia Simulation. *Anesthesia Analgesia* 71 (1990): 77–82.

Deinstadt, D.C. Dollars Plus Sense: Nine Considerations for Saving on Medical Equipment Maintenance. *Health Facilities Management* 15 (2002): 28–33.

Dhillon, B.S. Bibliography of Literature on Medical Equipment Reliability. *Microelectronics and Reliability* 20 (1980): 737–742.

Dhillon, B.S. *Design Reliability: Fundamentals and Applications.* Boca Raton, FL: CRC Press, 1999.

Dhillon, B.S. Human and Medical Device Reliability. In *Handbook of Reliability Engineering*, edited by H. Pham, 529–542. New York: Springer-Verlag Publishers, 2003.
Dhillon, B.S. Human Factors in Medical Devices. Proceedings of the 32nd International Conference on Computers and Industrial Engineering, 2003, 176–181.
Dhillon, B.S. Human Error in Medical Systems. Proceedings of the 6th ISSAT International Conference on Reliability and Quality in Design, 2000, 138–143.
Dhillon, B.S. Human Error in Medication. Proceedings of the 10th ISSAT International Conference on Reliability and Quality in Design, 2004, 325–329.
Dhillon, B.S. *Human Reliability and Error in Medical System*. River Edge, NJ: World Scientific Publishing, 2003.
Dhillon, B.S. *Medical Device Reliability and Associated Areas*. Boca Raton, FL: CRC Press, 2000.
Dhillon, B.S. Medical Device Software Failures. Proceedings of the Safety and Reliability International Conference, 2001, 113–123.
Dhillon, B.S. Methods for Performing Human Reliability and Error Analysis in Health Care. *International Journal of Health Quality Assurance* 16 (2003): 306–317.
Dhillon, B.S. Reliability Considerations in Medical Devices. Proceedings of the 43rd Congress of the European Organization for Quality Control, 1999, 400–418.
Dhillon, B.S. *Reliability Engineering Applications: Bibliography on Important Application Areas*. Gloucester, Ontario: Beta Publishers, 1992, Chapter 13.
Dhillon, B.S. *Reliability Engineering in Systems Design and Operation*. New York: Van Nostrand Reinhold, 1983, Chapter 11.
Dhillon, B.S. Reliability Technology for Manufacturers: Engineering Better Medical Devices. *Medical Device and Diagnostic Industry* 23 (2001): 94–99.
Dhillon, B.S. Reliability Technology in Health Care Systems. Proceedings of the IASTED International Symposium on Computers and Advanced Technology in Medicine, Healthcare, and Bioengineering, 1990, 84–87.
Dhillon, B.S. Risk Assessment in Medical Devices. Proceedings of the 6th International Conference on Probabilistic Safety Assessment and Management, 2002, 203–207.
Dhillon, B.S. Tools for Improving Medical Equipment Reliability and Safety. *Physics in Medicine and Biology* 39a (1994): 941.
Dhillon, B.S., Rajendran, M. Human Error in Health Care Systems. *International Journal of Reliability, Quality, and Safety Engineering* 10 (2003): 99–117.
Dhillon, B.S., Rajendran, M. Stochastic Analysis of Health Care Professional-Patient Interactions Subject to Human Error. Proceedings of the Fifth IASTED International Conference on Modelling, Simulation, and Optimization, 2005, 145–150.
Dickson, C. World Medical Electronics Market: An Overview. *Medical Devices & Diagnostic Industry* 6 (1984): 53–58.
Domizio, G.D., Davis, N.M., Cohen, M.R. The Growing Risk of Look-Alike Trademarks. *Medical Marketing & Media* May (1992): 24–30.
Donchin, Y., Biesky, M., Cotev, S., Gopher, D., Olin, M., Badihi, Y., Cohen, G. The Nature and Causes of Human Errors in a Medical Intensive Care Unit. Proceedings of the 33rd Annual Meeting of the Human Factors Society, 1989, 111–118.
Donchin, Y., Gopher, D., Olin, M., Badihi, Y. A Look into the Nature and Causes of Human Error in the Intensive Care Unit. *Critical Care Medicine* 23 (1995): 294–300.
Donohue, J., Apostolou, S.F. Shelf-Life Prediction for Radiation-Sterilized Plastic Devices. *Medical Device & Diagnostic Industry* 12 (1996): 124–129.
Doyal, L. The Mortality of Medical Mistakes, Doctors-Obligations and Patients-Expectation. *Practitioner* 231 (1987): 615–620.
Doyle, D.J. Do It By Design: Human Factors Issues in Medical Devices. *Canadian Journal of Anaesthesia* 47 (2000): 1259–1260.
Drefs, M.J. IEC 601-1 Electrical Safety Testing. *Compliance Engineering* XIV (1997): 11–14.

Drefs, M.J. IEC 601-1 Electrical Testing, Part 2: Performing the Tests. *Compliance Engineering* XIV (1997): 23–25.

Dripps, R.D., Lamont, A., Eckenhoff, J.E. The Role of Anesthesia in Surgical Mortality. *Journal of the American Medical Association* 178 (1991): 261–266.

Drury, C.G., Schiro, S.C., Czaja, S.J., Barnes, R.E. Human Reliability in Emergency Medical Response. Proceedings of the Annual Reliability and Maintainability Symposium, 1977, 38–42.

Dultgen, P., Meier, A. Requirements for Remote Maintenance for Medical Wearable Devices. Proceedings of the 7th Biennial Conference on Engineering Systems Design and Analysis, 2004, 367–374.

Dunn, J.D. Error in Medicine. *Annals of Internal Medicine* 134 (2001): 342–344.

Dutton, G. Do American Hospitals Get Away with Murders? *Business and Health* 18 (2000): 38–47.

Eberhard, D.P. Qualification of High Reliability Medical Grade Batteries. Proceedings of the Annual Reliability and Maintainability Symposium, 1989, 356–362.

Edwards, F.V. Before Design: Thoroughly Evaluate Your Concept. *Medical Device & Diagnostic Industry* 19 (1997): 46–50.

Egeberg, R.O. Engineers and the Medical Crisis. *Proceedings of the IEEE* 57 (1969): 1807–1808.

Eichhorn, J.H. Prevention of Intraoperative Anesthesia Accidents and Related Severe Injury Through Safety Monitoring. *Anesthesiology* 70 (1989): 572–577.

Eisele, G.R., Watkins, J.P. Survey of Workplace Violence in Department of Energy Facilities. Report, U.S. Department of Energy, Office of Occupational Medicine and Medical Surveillance, Washington, DC, December 1995.

Eiselstem, L.E., James, B. Medical Device Failures: Can We Learn From Our Mistakes? Proceedings of the Materials and Processes for Medical Devices Conference, 2004, 3–11.

Eisenberg, J.M. Medical Error Accountability. *Family Practice News* 30 (2000): 12.

Eisenberg, J.M. The Best Offence Is a Good Defense Against Medical Errors: Putting the Full-Court Press on Medical Errors. Agency for Healthcare Research and Quality, Duke University Clinical Research Institute, Rockville, MD: January 20, 2000.

Elahi, B.J. Safety and Hazard Analysis for Software-Controlled Medical Devices. Proceedings of the 6th Annual IEEE Symposium on Computer-Based Medical Systems, 1993, 10–15.

Elkin, P.L., Brown, S.H., Carter, J. Guideline and Quality Indicators for Development, Purchase and Use of Controlled Health Vocabularies. *International Journal of Medical Informatics* 68 (2002): 175–186.

Elliott, L., Mojdehbakhsh, R. A Process for Developing Safe Software. Proceedings of the 7th Symposium on Computer-Based Medical Systems, 1994, 241–246.

Epstein, R.M. Morbidity and Mortality from Anesthesia. *Anesthesiology* 49 (1978): 388–389.

Espinosa, J. A., Nolan, T.W. Reducing Errors Made by Emergency Physician in Interpreting Radiographs: Longitudinal Study. *British Medical Journal* 320 (2000): 737–740.

Estrin, N.F., ed. *The Medical Device Industry: Science, Technology, and Regulation in a Competitive Environment.* New York: Marcel Dekker, 1990.

Fairhurst, G.A., Murphy, K.L. Help Wanted. Proceedings of the Annual Reliability and Maintainability Symposium, 1976, 103–106.

Famularo, G., Salvani, P., Terranova, A. Clinical Errors in Emergency Medicine: Experience at the Emergency Department of an Italian Teaching Hospital. *Academic Emergency Medicine* 7 (2000): 1278–1281.

Feinstein, A.R. Errors in Getting and Interpreting Evidence. *Perspectives in Biology and Medicine* 41 (1999): 45–58.

Feldman, S.E., Roblin, D.W. Medical Accidents in Hospitals Care: Applications of Failure Analysis to Hospitals Quality Appraisal. *Journal of Quality Improvement* 23 (1997): 567–580.

Ferman, J. Medical Errors Spur New Patient-Safety Measures. *Healthcare Executive* March (2000): 55–56.

Ferner, R.E. Medication Errors that Have Led to Manslaughter Charges. *British Medical Journal* 321 (2000): 1212–1216.

Ferner, R. E. Reporting of Critical Incidents Shows How Errors Occur. *British Medical Journal* 311 (1995): 1368.

Finfer, S.R. Pacemaker Failure on Induction of Anaethesia. *British Journal of Anaesthesia* 66 (1991): 509–512.

Finger, T.A. The Alpha System. Proceedings of the Annual Reliability and Maintainability Symposium, 1976, 92–96.

Fisher, M.J. Eliminating Medical Errors Is Sound Risk Management. *National Underwriter*, October 5, 1998, 51.

Fletcher, M. Safety-Engineered Medical Devices Become Mandatory in Saskatchewan. *Canadian Nurse* 101 (2005): 12–13.

Flickinger, J.C., Lunsford, L.D. Potential Human Error in Setting Stereo Tactic Coordinates for Radiotherapy: Implication for Quality Assurance. *International Journal of Radiation Oncology, Biology, Physics* 27 (1993): 397–401.

Food and Drug Administration. Medical Devices; Current Good Manufacturing Practice (GMP) Final Rule; Quality System Regulation. Washington, DC: Food and Drug Administration, Department of Health and Human Services, 1996.

Food and Drug Administration. Preproduction Quality Assurance Planning Recommendations for Medical Device Manufacturers, Compliance Guidance Series. HHS Publication FDA 90–4236. Rockville, MD: Food and Drug Administration, September 1989.

Food and Drug Administration. Statistical Guidance for Clinical Trials of Non-Diagnostic Medical Devices. Rockville, MD: Center for Devices and Radiological Health, Food and Drug Administration, 1996.

Fox, G.N. Minimizing Prescribing Errors in Infants and Children. *American Family Physician* 53 (1996): 1319–1325.

Fried, R.A. TQM in the Medical School: A Report from the Field. Proceedings of the 45th Annual Quality Congress Transactions, 1991, 113–118.

Fries, R. Human Factors and System Reliability. *Medical Device Technology* March (1992): 42–46.

Fries, R.C. *Medical Device Quality Assurance and Regulatory Compliance*. New York: Marcel Dekker, 1998.

Fries, R.C. *Reliability Assurance for Medical Devices, Equipment, and Software*. Buffalo Grove, IL: Interpharm Press, 1991.

Fries, R.C. *Reliable Design of Medical Devices*. New York: Marcel Dekker, 1997.

Fries, R.C., Pienkowski, P., Jorgens, J. Safe, Effective and Reliable Software Design and Development for Medical Devices. *Medical Instrumentation* 30 (1996): 75–80.

Fries, R.C., Willingmyre, G.T., Simons, D., Schwartz, R.T. Software Regulation. In *The Medical Device Industry: Science, Technology, and Regulation in a Competitive Environment*, edited by N.F. Estrin, 557–569. New York: Marcel Dekker, 1990.

Frings, G.W., Grant, L. Who Moved My Sigma... Effective Implementation of the Six Sigma Methodology to Hospitals. *Quality and Reliability Engineering International* 21 (2005): 311–328.

Gaba, D.M. Human Error in Anesthetic Mishaps. *International Anesthesiology Clinics* 27 (1989): 137–147.

Gaba, D.M. Human Performance Issues in Anaesthesia Patient Safety. *Problems in Anaesthesia* 5 (1991): 329–330.

Gaba, D.M., Howard S.K. Conference on Human Error in Anesthesia (meeting report). *Anesthesiology* 75 (1991): 553–554.

Gaba, D.M., Maxwell, M., DeAnda, A. Anesthetic Mishaps: Breaking the Chain of Accident Evolution. *Anesthesiology* 66 (1987): 670–676.

Gawande, A.A., Thomas, E.J., Zinner, M.J. The Incidence and Nature of Surgical Adverse Events in Colorado and Utah in 1992. *Surgery* 126 (1999): 66-75.

Gebhart, F. Hospitals—VHA and VA Alike—Launch Programs to Cut Errors. *Drug Topics* 20 (2000): 53.

Gechman, R. Tiny Flaws in Medical Design Can Kill. *Hospital Topics* 46 (1968): 23–24.

Gehlot, V., Sloane, E.B. Ensuring Patient Safety in Wireless Medical Device Networks. *Computer* 39 (2006): 54–60.

Gerlin, A. Medical Mistakes. *The Inquirer,* September 12, 1999, 65–64.

Ghahramani, B. An Internet Based Total Quality Management System. Proceedings of the 34th Annual Meeting of the Decision Sciences Institute, 2003, 345–349.

Gingerich, D., ed. *Medical Product Liability.* New York: F&S Press, 1981.

Ginsburg, G. Human Factors Engineering: A Tool for Medical Evaluation in Hospital Procurement Decision-Making. *Journal of Biomedical Informatics* 38 (2005): 213–219.

Giuntini, R.E. Developing Safe, Reliable, Medical Devices. *Medical Device & Diagnostic Industry* 35 (2000): 10–15.

Glick, T.H., Workman, T.P., Gaufrerg, S.V. Human Error in Medicine. *Academic Emergency Medicine* 7 (2000): 1272–1277.

Goodman, G.R. Medical Device Error. *Critical Care Nursing Clinics of North America* 14 (2002): 407–416.

Gopher, D., Olin, M., Badihi, Y. The Nature and Causes of Human Errors in a Medical Intensive Care Unit. Proceedings of the Human Factors Society 33rd Annual Meeting, 1989, 956–960.

Gosbee, J. The Discovery Phase of Medical Device Design: A Blend of Intuition, Creativity and Science. *Medical Device & Diagnostic Industry* 19 (1997): 79–85.

Gosbee, J., Ritchie, E.M. Human Computer Interaction and Medical Software Development. *Interactions* 4 (1997): 13–18.

Grabarz, D.F., Cole, M..F. Developing a Recall Program. In *The Medical Device Industry: Science, Technology, and Regulation in a Competitive Environment*, edited by N.F. Estrin, 335–351. New York: Marcel Dekker, 1990.

Graham, R. Why Are We Not Teaching Health Care Professionals About Human Error? *Hospital Pharmacist* 13 (2006): 371–372.

Grant, L.J. Product Liability Aspects of Bioengineering. *Journal of Biomedical Engineering* 12 (1990): 262–266.

Grant, L.J. Regulations and Safety in Medical Equipment Design. *Anaethesia* 53 (1998): 1–3.

Grasha, A.F. Into the Abyss: Seven Principles for Identifying the Causes of and Preventing Human Error in Complex Systems. *American Journal of Health-System Pharmacy* 57 (2000): 554–564.

Gravenstein, J.S. How Does Human Error Affect Safety in Anesthesia? *Surgical Oncology Clinics of North America* 9 (2000): 81–95.

Green, R. The Psychology of Human Error. *European Journal of Anaesthesiology* 16 (1999): 148–55.

Gruppen, L.D., Wolf, F.M., Billi, J.E. Information Gathering and Integration as Sources of Error in Diagnostic Decision Making. *Medical Decision Making* 11 (1991): 233–239.

Gunn, I.P. Patient Safety and Human Error: The Big Picture. *CRNA: The Clinical Forum for Nurse Anesthetists* 11 (2000): 41–48.

Gurwitz, J.H., Sanchez-Cross, M.T., Eckler, M.A., Matulis, J. The Epidemiology of Adverse and Unexpected Events in the Long-Term Care Setting. *Journal of the American Geriatric Society* 42 (1994): 33–38.

Guyton, B. Human Factors and Medical Devices: A Clinical Engineering Perspective. *Journal of Clinical Engineering* 27 (2002): 116–122.

Hales, R.F. Quality Function Deployment in Concurrent Product/Process Development. Proceedings of the 6th Annual IEEE Symposium on Computer-Based Medical Systems, 1993, 28–33.

Hallam, K. An Erosion of Trust: Survey Finds Consumers Fear Medical Errors and Want Better Protection From Such Mistakes. *Modern Healthcare* 30 (2000): 30–31.

Handler, J.A., Gillam, M., Sanders, A.B. Defining, Identifying, and Measuring Medical Error in Emergency Medicine. *Academic Emergency Medicine* 7 (2000): 1183–1188.

Hart G.K., Baidwin, I., Ford, J., Gutteridge, G.A. Adverse Incidents in Intensive Care Survey. *Anaesthesia and Intensive Care* 21 (1993): 65–68.

Haslam, K.R., Bruner, J.M.R. The Epidemiology of Failure in Cardiac Monitoring Systems. *Medical Instrumentation* 7 (1973): 293–296.

Hatlie, M. J. Scapegoating Won't Reduce Medical Errors. *Medical Economics* 77 (2000): 92–100.

Haugh, R. Medical/Surgical Equipment: All Eyes on Safety. *Hospitals and Health Networks* 75 (2001): 30–33.

Hearnshaw, H. Human Error in Medicine. *Ergonomics* 39 (1996): 899–900.

Hector, J., Tom, P. Missed Diagnosis of Acute Cardiac Ischemia in the Emergency Department. *New England Journal of Medicine* 342 (2000): 1163–1170.

Helmreich, R.L., Schaefer, H.G. Team Performance in the Operating Room. In *Human Error in Medicine*, edited by M.S. Bogner, 225–254. Hillsdale, NJ: Lawrence Erlbaum Associates, 1994.

Heydrick, L., Jones, K.A., Applying Reliability Engineering During Product Development. *Medical Device & Diagnostic Industry* 18 (1996): 80–84.

Hobgood, C.D., John, O., Swart, L. Emergency Medicine Resident Errors: Identification and Educational Utilization. *Academic Emergency Medicine* 7 (2000): 1317–1320.

Hogl, O., Muller, M., Stoyan, H. On Supporting Medical Quality with Intelligent Data Mining. Proceedings of the 34th IEEE Annual Hawaii International Conference on System Sciences, 2001, 141–145.

Hollnagel, E. *Cognitive Reliability and Error Analysis.* Amsterdam: Elsevier, 1998.

Hooten, W.F. A Brief History of FDA Good Manufacturing Practices. *Medical Device & Diagnostic Industry* 18 (1996): 96.

Hopps, J.A. Electrical Hazards in Hospital Instrumentation. Proceedings of the Annual Symposium on Reliability, 1969, 303–307.

Horan, S. Is Human Error the Cause of Accident and Disaster? *Occupational Health* 45 (1993): 169–172.

Hough, G.W. *Preproduction Quality Assurance for Healthcare Manufacturers.* Buffalo Grove, IL: Interpharm Press, 1997.

Hughes, C.M., Honig, P., Phillips, J., Woodcook, J., Richard E. Anderson, R.E., et al. How Many Deaths Are Due to Medical Errors? *Journal of the American Medical Association* 284 (2000): 2187–2189.

Hupert, N., Lawthers, A.G., Brennan, T.A., Peterson, L.M. Processing the Tort Deterrent Signal: A Qualitative Study. *Social Science Medicine* 43 (1996): 1–11.

Hutt, P.B. A History of Government Regulation of Adulteration and Misbranding of Medical Devices. In *The Medical Device Industry: Science, Technology, and Regulation in a Competitive Environment*, edited by N.F. Estrin, 17–33. New York: Marcel Dekker, 1990.

Hyman, W.A. Errors in the Use of Medical Equipment. In *Human Error in Medicine*, edited by M.S. Bogner, 327–347. Hillsdale, NJ: Lawrence Erlbaum Associates, 1994.

Ibojie, J., Urbaniak, S.J. Comparing Near Misses with Actual Mistransfusion Events: A More Accurate Reflection of Transfusion Errors. *British Journal of Haematology* 108 (2000): 458–460.

International Electrotechnical Commission. Safety of Medical Electrical Equipment, Part 1: General Requirements. IEC 601-1. Geneva, Switzerland: International Electrotechnical Commission, 1977.

International Organization for Standardization. Medical Devices-Risk Management—Part I: Application of Risk Analysis to Medical Devices. ISO/DIS 14971. Geneva, Switzerland: International Organization for Standardization (ISO), 1996.

Iseli, K.S., Burger, S. Diagnostic Errors in Three Medical Eras: A Necropsy Study. *Lancet* 355 (2000): 2027–2031.

Israelski, E.W., Muto, W.H. Human Factors Risk Management as a Way to Improve Medical Device Safety; A Case Study of the Therac 25 Radiation Therapy System. *Joint Commission Journal of Quality and Safety* 30 (2004): 689–695.

Ivy, M.E., Cohn, K. Human Error in Hospitals and Industrial Accidents. *Journal of the American College of Surgeons* 192 (2001): 421.

Iz, P.H., Warren, J., Sokol, L. Data Mining for Healthcare Quality, Efficiency, and Practice Support. Proceedings of the IEEE 34th Annual Hawaii International Conference on System Sciences, 2001, 147–152.

Jackson, T. How the Media Report Medical Errors. *British Medical Journal* 322 (2001): 562–564.

Jacky, J. Risks in Medical Electronics. *Communications of the ACM* 33 (1990): 138.

James, B.C., Hammond, E.H. The Challenge of Variation in Medical Practice. *Archives of Pathology and Laboratory Medicine* 124 (2000): 1001–1004.

Jameson, F.L. Exposing Medical Mistakes. *New York Times,* February 24, 2000, A4.

Johns, R.J. What is Blunting the Impact of Engineering on Hospitals? *Proceedings of the IEEE* 57 (1969): 1823–1827.

Johnson, J.P. Reliability of ECG Instrumentation in a Hospital. Proceedings of the Annual Symposium on Reliability, 1969, 314–318.

Johnson, W.G., Brennan, T.A., Newhouse, J.P., Leape, L.L., Lawthers, A.G., Hiath, H.H., Weiler, P.C. The Economic Consequences of Medical Injuries. *Journal of the American Medical Association* 267 (1992): 2487–2492.

Joice, P., Hanna, G.B., Cuschieri, A. Errors Enacted During Endoscopic Surgery—A Human Reliability Analysis. *Applied Ergonomics* 29 (1998): 409–415.

Jones, J., Hunter D. Qualitative Research: Consensus methods for medical and health services research. *British Medical Journal* 311 (1995): 376–380.

Jorgens, J. Computer Hardware and Software as Medical Devices. *Medical Device & Diagnostic Industry* 5 (1983): 62–67.

Joyce, E. Malfunction 54: Unraveling Deadly Medical Mystery of Computerized Accelerator Gone Awry. *American Medical News,* October 3, 1986, 1–8.

Kagey, K.S. Reliability in Hospital Instrumentation. Proceedings of the Annual Reliability and Maintainability Symposium, 1973, 85–88.

Kandel, G.L., Ostrander, L.E. Preparation of the Medical Equipment Designer: Academic Opportunities and Constraints. Proceedings of the First Symposium on Human Factors in Medical Devices, 1989, 11–13.

Karp, D. Your Medication Orders Can Become Malpractice Traps. *Medical Economics*, February 1, 1988, 79–91.

Kaye, R. Human Factors in Medical Device Use Safety: How to Meet the New Challenges. Proceedings of the Human Factors and Ergonomics Society Meeting, 2002, 550–552.

Keenan, R.L., Boyan, P. Cardiac Arrest Due to Anesthesia. *Journal of the American Medical Association* 253 (1985): 2373–2377.

Keller, A.Z., Kamath, A.R.R., Peacock, S.T. A Proposed Methodology for Assessment of Reliability, Maintainability and Availability of Medical Equipment. *Reliability Engineering* 9 (1984): 153–174.

Keller, J. Top Ten Safety Issues with Medical Devices. *OR Manager* 19 (2003): 24–25.

Keller, J.P. Human Factors Issues in Surgical Devices. Proceeding of the First Symposium on Human Factors in Medical Devices, 1989, 34–36.

Kelley, J.B. Medical Diagnostic Device Reliability Improvement and Prediction Tools: Lessons Learned. Proceedings of the Annual Reliability and Maintainability Symposium, 1999, 29–31.

Keselman, A., Tang, X., Patel, V.L., Johnson, T.R., Zhang, J. Institutional Decision-Making for Medical Purchasing: Evaluating Patient Safety. *Medinfo* 11 (2004): 1357–1361.

Khorramshahgol, R., Al-Barmil, J., Stallings, R.W. TQM in Hospitals and the Role of Information Systems in Providing Quality Services. Proceedings of the IEEE Annual International Engineering Management Conference, 1995, 196–199.

Kieffer, R.G. Validation and the human element. *PDA Journal of Pharmaceutical Science and Technology* 52 (1998): 52–54.

Kim, J.S. Determining Sample Size for Testing Equivalence. *Medical Device & Diagnostic Industry* 19 (1997): 114–117.

Kim, J.S., Larsen, M. Improving Quality with Integrated Statistical Tools. *Medical Device & Diagnostic Industry* 18 (1996): 78–82.

Kimberly, J.R., Minvielle, E. *The Quality Imperative: Measurement and Management of Quality in Healthcare.* London: Imperial College Press, 2000.

King, R. Six Sigma and Its Application in Healthcare. Proceedings of the 57th Annual Quality Congress: Expanding Horizons, 2003, 39–47.

Kirk, R. *Healthcare Quality & Productivity: Practical Management Tool.* Rockville, MD: Aspen Publishers, 1988.

Kirwan, B. Human Error Identification Techniques for Risk Assessment of High Risk System, Part 1: Review and Evaluation of Techniques. *Applied Ergonomics* 29 (1998): 157–178.

Kirwan, B. Human Error Identification Techniques for Risk Assessment of High Risk System, Part 2: Towards a Framework Approach. *Applied Ergonomics* 29 (1998): 299–318.

Klatzky, R.L., Geiwitz, J., Fischer, S. Using Statistics in Clinical Practice: A Gap Between Training and Application. In *Human Error in Medicine*, edited by M.S. Bogner, 123–140. Hillsdale, NJ: Lawrence Erlbaum Associates, 1994.

Klemola, U.M. The Psychology of Human Error Revisited. *European Journal of Anaesthesiology* 17 (2000): 401.

Knaus, W.A., Draper, E.A., Wagner, D.P. An Evaluation of Outcome from Intensive Care in Major Medical Centres. *Annals of Internal Medicine* 104 (1986): 410–418.

Knepell, P.L. Integrating Risk Management with Design Control. *Medical Device & Diagnostic Industry* 20 (1998): 83–89.

Knight, J.C. Issues of Software Reliability in Medical Systems. Proceedings of the 3rd Annual IEEE Symposium on Computer-Based Medical Systems, 1990, 124–129.

Knuckey, L.M. EPSM 2005 Workshop IEC 6060-1: Medical Electrical Equipment—Part I: General Requirements for Basic Safety and Essential Performance. *Australian Physical and Engineering Sciences in Medicine* 29 (2006): 117–118.

Kohn, L. T., Corrigan, J. M., Donaldson, M.S., eds. *To Err Is Human: Building a Safer Health System.* Washington, DC: Institute of Medicine, National Academy of Medicine, National Academies Press, 1999.

Kortstra, J.R.A. Designing for the User. *Medical Device Technology* January/February (1995): 22–28.

Krueger, G.P. Fatigue, Performance, and Medical Error. In *Human Error in Medicine*, edited by M.S. Bogner, 311–326. Hillsdale, NJ: Lawrence Erlbaum Associates, 1994.

Kruk, S.M. Do Diligence: A Look at The 10 Most Common Errors in Medical Equipment Maintenance Invoices. *Health Facilities Management* 18 (2005): 34–36.

Kumar, V., Barcellos, W.A., Mehta, M.P. An Analysis of Critical Events in a Teaching Department for Quality Assurance: A Survey of Mishaps During Anesthesia. *Anesthesia* 43 (1988): 879–883.

Kuwahara, S.S. *Quality Systems and GMP Regulations for Device Manufacturers.* Buffalo Grove, IL: Interpharm Press, 1998.

Kyriacou, D.N., Coben, J.N. Errors in Emergency Medicine: Research Strategies. *Academic Emergency Medicine* 7 (2000): 1201–1203.

Labovitz, G.H. Total-quality Health Care Revolution. *Quality Progress* 24 (1991): 45–47.

Lane, R. Applying Hierarchical Task Analysis to Medication Administration Errors. *Applied Ergonomics* 37 (2006): 669–679.

Larimer, M.L., Bergquist, T.M. A Comparison of Three Quality Improvement Systems for Health Care. Proceedings of the 34th Annual Meeting of the Decision Sciences Institute, 2003, 1843–1848.

Le Cocq, A.D. Application of Human Factors Engineering in Medical Product Design. *Journal of Clinical Engineering* 12 (1987): 271–277.

Le Duff, F., Daniel, S., Kamendje, B. Monitoring Incident Report in the Healthcare Process to Improve Quality in Hospitals. *International Journal of Medical Informatics* 74 (2005): 111–117.

Leape, L.L. Error in Medicine. *Journal of the American Medical Association* 272 (1994): 1851–1857.

Leape, L.L. The Preventability of Medical Injury. In *Human Error in Medicine*, edited by M.S. Bogner, 13–27. Hillsdale, NJ: Lawrence Erlbaum Associates, 1994.

Leape, L.L., Cullen, D.J., Laird, N., Petersen, L.A., Teich, J.M., Burdick, E., et al. Effect of Computerized Physician Order Entry and a Team Intervention on Prevention of Serious Medication Errors. *Journal of the American Medical Association* 280 (1998): 1311–1316.

Leape L.L., Gullen, D.J., Clapps, M.D. Pharmacist Participation on Physician Rounds and Adverse Drug Events in the Intensive Care Unit. *Journal of the American Medical Association* 282 (1999): 267–270.

Leape, L.L., Swankin, D.S., Yessian, M.R. A Conversation on Medical Injury. *Public Health Reports* 114 (1999): 302–317.

Leape, L.L, Woods, D.D., Hatlie, M.J., Kizre, K.W. Promoting Patient Safety by Preventing Medical Error. *Journal of the American Medical Association* 280 (1998): 1444–1448.

Ledley, R.S. Practical Problems in the Use of Computers in Medical Diagnosis. *Proceedings of the IEEE* 57 (1969): 1941.

Leffingwell, D.A., Norman, B. Software Quality in Medical Devices: A Top-Down Approach. Proceedings of the 6th Annual IEEE Symposium on Computer-Based Medical Systems, 1993, 307–311.

Levkoff, B. Increasing Safety in Medical Device Software. *Medical Device & Diagnostic Industry* 18 (1996): 92–97.

Lin, L., Isla, R., Doniz, K., Harkness, H., Vicente, K.J., Doyle, D.J. Applying Human Factors to the Design of Medical Equipment: Patient Controlled Analgesia. *Journal of Clinical Monitoring and Computing* 14 (1998): 253–263.

Lin, L., Vicente, K.J., Doyle, D.J. Patient Safety, Potential Adverse Drug Events, and Medical Device Design: A Human Factors Approach. *Journal of Biomedical Informatics* 34 (2001): 274–284.

Linberg, K.R. Defining the Role of Software Quality Assurance in a Medical Device Company. Proceedings of the 6th Annual IEEE Symposium on Computer-Based Medical Systems, 1993, 278–283.

Ling, M.T.K, Sandford, C., Sadik, A., Blom, H., Ding, S.Y., Woo, L. How Processing Conditions Can Cause Degradation and Failure in Medical Devices. *Plastics Engineering* 57 (2001): 50–55.

Link, D.M. Current Regulatory Aspects of Medical Devices. Proceedings of the Annual Reliability and Maintainability Symposium, 1972, 249–250.

Lo, M. Formulating Equipment Maintenance Strategy in Hospital Authority (HA). IEE 3rd Seminar on Appropriate Medical Technology for Developing Countries. IEE Colloquium (Digest) No. V4-10408, 2004, 61–64.

Lori, A.B., Carol, S. An Alternative Strategy for Studying Adverse Events in Medical Care. *Lancet* 349 (1997): 309–313.

Ludbrook, G.L., Webb, R.K., Fox, M.A., Singletons, R.J. Problems Before Induction of Anaesthesia: An Analysis of 2000 Incidents Report. *Anaesthesia and Intensive Care* 21 (1993): 593–595.

Lunn, J., Devlin, H. Lessons From the Confidential Inquiry into Preoperative Deaths in Three NHS Regions. *Lancet* 2 (1987): 1384–1386.

Maddox, M.E. Designing Medical Devices to Minimize Human Error. *Medical Device & Diagnostic Industry* 19 (1997): 160–180.

Marconi, M., Sirchia, G. Increasing Transfusion Safety by Reducing Human Error. *Current Opinion in Hematology* 7 (2000): 382–386.

Marsden, P., Hollnagel, E. Human Interaction With Technology: The Accidental User. *Acta Psychologica* 91 (1996): 345–358.

Marshall, D. Health Care Quality: It Is Not "Job One"! Proceedings of the 56th ASQ Annual Quality Congress, 2002, 83–91.

Marszalek-Gaucher, E. *Total Quality in Healthcare: From Theory to Practice.* San Francisco: Jossey-Bass Publishers, 1993.

Maschino, S., Duffell, W. Can Inspection Time be Reduced?: Developing a HACCP Plan. *Medical Device & Diagnostic Industry* 20, 10 (1998): 64– 68.

Mayor, S. English NHS to Set Up New Reporting System for Errors. *British Medical Journal* 320 (2000): 1689.

Mazur, G. Quality Function Development for a Medical Device. Proceedings of the 6th Annual IEEE Symposium on Computer-Based Medical Systems, 1993, 22–27.

McCadden, P. Medical Errors. Proceedings of a Conference on Enhancing Patient Safety and Reducing Errors in Health Care, National Patient Safety Foundation, Rancho Mirage, CA, 1999, 68–69.

McClure, M.L. Human Error—A Professional Dilemma. *Journal of Professional Nursing* 7 (1991): 207.

McConnel, E.A. Pointed Strategies for Needle Stick Prevention. *Nursing Management* 30 (1999): 57–60.

McCray, S. *Medical Errors in the US Medical System, Institute of Medicine.* Washington, DC: Institute of Medicine, National Academy of Medicine, National Academies Press, 1999.

McCune, C.G. Development of an Automated Pacemaker Testing System. Proceedings of the 27th Midwest Symposium on Circuits and Systems, 1984, 9–12.

McDaniel, J.G. Improving System Quality Through Software Evaluation. *Computers in Biology and Medicine* 32 (2002): 127–140.

McLinn, J.A. Reliability Development and Improvement of a Medical Instrument. Proceedings of the Annual Reliability and Maintainability Symposium, 1996, 236–242.

McManus, B. A Move to Electronic Patient Records in the Community: A Qualitative Case Study of a Clinical Data Collection System. Problems Caused by Inattention to Users and Human Error. *Topics in Health Information Management* 20 (2000): 23–37.

McManus, J. Hong Kong Issues Guidelines to Prevent Further Medical Blunders. *British Medical Journal* 315 (1997): 967–972.

Mead, D. Human Factors Evaluation of Medical Devices. Proceedings of the First Symposium on Human Factors in Medical Devices, 1989, 17–18.

Meadows, S. Human Factors Issues with Home Care Devices. Proceedings of the First Symposium on Human Factors in Medical Devices, 1989, 37–38.

Meguro, T. Minimizing Human Error Based on Clinical Engineering and Safety Management of Medical Equipment. *IRYO Japanese Journal of National Medical Services* 59 (2005): 34–35.

Mehra, R.H., Eagle, K. A. Missed Diagnosis of Acute Coronary Syndromes in the Emergency Room-Continuing Challenges. *New England Journal of Medicine* 342 (2000): 1207–1209.

Merry, M.D. Healthcare's Need for Revolutionary Change. *Quality Progress* 36 (2003): 31–35.

Meyer, J.L. Some Instrument Induced Errors in the Electrocardiogram. *Journal of the American Medical Association* 201 (1967): 351–358.

Micco, L.A. Motivation for the Biomedical Instrument Manufacturer. Proceedings of the Annual Reliability and Maintainability Symposium, 1972, 242–244.

Miller, M.J., Bridgeman, E.A. Role of AAMI in the Voluntary Standards Process. In *The Medical Device Industry: Science, Technology, and Regulation in a Competitive Environment*, edited by N.F. Estrin, 163–175. New York: Marcel Dekker, 1990.

Moeller, J., O'Reilly, J.B., Elser, J. Quality Management in German Health Care—The EFQM Excellence Model. *International Journal of Health Care Quality Assurance* 13 (2000): 254–258.

Moliver, M. Computerization Helps Reduce Human Error in Patient Dosing. *Contemporary Long Term Care* 10 (1987): 56–58.

Montanez, J. *Medical Device Quality Assurance Manual*. Buffalo Grove, IL: Interpharm Press, 1996.

Moraes, L., Garcia, R. Contribution for the Functionality and the Safety in Magnetic Resonance: An Approach for the Imaging Quality. Proceedings of the 25th Annual International Conference of the IEEE Engineering in Medicine and Biology Society, 2003, 3613–3616.

Moray, D. Error Reduction as a Systems Problem. In *Human Error in Medicine*, edited by M.S. Bogner, 67–92. Hillsdale, NJ: Lawrence Erlbaum Associates, 1994.

Mosenkis, R. Critical Alarms. Proceedings of the First Symposium on Human Factors in Medical Devices, 1989, 25–27.

Muldur, S. Computer-Aided Planned Maintenance System for Medical Equipment. *Journal of Medical Systems* 27 (2003): 393–398.

Murray, K. Canada's Medical Device Industry Faces Cost Pressures, Regulatory Reform. *Medical Device & Diagnostic Industry* 19 (1997): 30–39.

Nakata, Y., Fujiwara, M.O., Goto, T. Risk Attitudes of Anesthesiologists and Surgeons in Clinical Decision Making with Expected Years of Life. *Journal of Clinical Anesthesia* 12 (2000): 146–150.

Narumi, J. Analysis of Human Error in Nursing Care, *Accident Analysis and Prevention* 31 (1999): 625–629.

National Bureau of Standards. Reliability Technology for Cardiac Pacemakers. NBS/FDA Workshop, June 1974. NBS Publication No. 400–28. Washington, DC: National Bureau of Standards, Department of Commerce, 1974.

Naylor, D. Reporting Medical Mistakes and Misconduct. *Canadian Medical Association Journal* 160 (1999): 1323–1324.

Neumann, P.G. Some Computer-Related Disasters and Other Egregious Horrors. Proceedings of the 7th Annual Conference of the IEEE/Engineering in Medicine and Biology Society, 1985, 1238–1239.

Nevland, J.G. Electrical Shock and Reliability Considerations in Clinical Instruments. Proceedings of the Annual Symposium on Reliability, 1969, 308–313.

Newhall, C. The Institute of Medicine Report on Medical Errors. *New England Journal of Medicine* 343 (2000): 105–109.
Nichols, T.R., Dummer, S. Assessing Pass/Fail Testing When There Are No Failures to Assess. *Medical Device & Diagnostic Industry* 19 (1997): 97–100.
No author. Tracking Medical Errors, From Humans to Machines. *New York Times*, March 31, 1992, 81.
Nobel, J.J. Medical Device Accident Reporting: Does It Improve Patient Safety? *Studies in Health Technology and Informatics* 28 (1996): 29–35.
Nobel, J.J. Medical Device Failures and Adverse Effects. *Pediatric Emergency Care* 7 (1991): 120–123.
Nolan, T.W. System Changes to Improve Patient Safety. *British Medical Journal* 320 (2000): 771–773.
Nordenberg, T. Make No Mistakes: Medical Errors Can Be Deadly Serious. *FDA Consumer* September–October (2000): 104–106.
Norman, J.C., Goodman, L. Acquaintance with and Maintenance of Biomedical Instrumentation. *Journal of the Association of Advanced Medical Instrumentation* September (1966): 8–9.
Northrup, S.J. Reprocessing Single-Use Devices: An Undue Risk. *Medical Device & Diagnostic Industry* 21 (1999): 38–41.
Northup, S.J. Safety Evaluation of Medical Devices: U.S. Food and Drug Administration and International Standards Organization Guidelines. *International Journal of Toxicology* 18 (1999): 275–283.
Norton, G.S., Romlein, J., Lyche, D.K. PACS 2000 – Quality Control using the Task Allocation Chart. Proceedings of the SPIE Medical Imaging Conference, 2000, 384–389.
O'Leary, D.S. Accreditation's Role in Reducing Medical Errors. *British Medical Journal* 320 (2000): 727–728.
O'Reilly, M.V., Murnaghan, D.P., Williams, M.B. Transvenous Pacemaker Failure Induced by Hyperkalemia. *Journal of the American Medical Association* 228 (1974): 336–337.
O'Rourke, K. The Human Factors in Medical Error. *New England Journal of Medicine* 304 (1981): 634–641.
O'Shea, A. Factors Contributing to Medication Errors: A Literature Review. *Journal of Clinical Nursing* 8 (1999): 496–504.
Offredy, M., Scott, J., Moore, R. Improving Safety and Learning: Case Study of an Incident Involving Medical Equipment. *Quality in Primary Care* 12 (2004): 19–27.
Okeyo, T. M., Adelhardt, M. *Health Professionals Handbook on Quality Management in Healthcare in Kenya.* Centre for Quality in Healthcare, Nairobi, Kenya, 2003.
Okuyama, Y., Sakamoto, K. Study of Failure Analysis for Safety Security During Medical Care. *Transactions of the Institute of Electronics, Information, and Communication Engineers* J88A (2005): 511–518.
Oliver, S. The Healthy Standard. *Quality World* 29 (2003): 16–20.
Olivier, D.P. Engineering Process Improvement Through Error Analysis. *Medical Device & Diagnostic Industry* 21 (1999): 130–136.
Ovretveit, J. System Negligence Is at the Root of Medical Error. *International Journal of Health Care Quality Assurance* 13 (2000): 103–105.
Owei, V. Healthcare Quality and Productivity: Framework for an Information Technology Impact Architecture. Proceedings of the Annual Meeting of the Decision Sciences Institute, 1998, 46–48.
Ozog, H. Risk Management in Medical Device Design. *Medical Device & Diagnostic Industry* 19 (1997): 112–115.

Palmer, P., Mason, L., Dunn, M. A Case Study in Healthcare Quality Management: A Practical Methodology for Auditing Total Patient X-ray Dose During a Diagnostic Procedure. Proceedings of the 7th Biennial Conference on Engineering Systems Design and Analysis, 2004, 469–474.

Passey, R.D. Foresight Begins with FMEA. Delivering Accurate Risk Assessments. *Medical Device Technology* 10 (1999): 88–92.

Pelnik, T.M., Suddarth, G.J. Implementing Training Programs for Software Quality Assurance Engineers. *Medical Device & Diagnostic Industry* 20 (1998): 75–85.

Perper, J.A. Life-Threatening and Fatal Therapeutic Misadvantages. In *Human Error in Medicine*, edited by M.S. Bogner, 27–52. Hillsdale, NJ: Lawrence Erlbaum Associates, 1994.

Persson, J., Ekberg, K., Linden, M. Work Organization, Work Environment and the Use of Medical Equipment: A Survey Study of the Impact on Quality and Safety. *Medical & Biological Engineering & Computing* 31 (1993): 20–24.

Peters, M. Implementation of Rules Based Computerized Bedside Prescribing and Administration: Intervention Study. *British Medical Journal* 320 (2000): 750–753.

Peterson, M.G.E. The Probability of Failure Depends on Who is Asking. Proceedings of the 8th IEEE Symposium on Computer Based-Medical Systems, 1995, 51–56.

Phillips, D.P., Christenfeld, N., Glynn, L.M. Increase in US Medication-Error Deaths Between 1983–1993. *Lancet* 1351 (1998): 1024–1029.

Pickett, R.M., Triggs, T.J. *Human Factors in Health Care.* Lexington, MA: Lexington Books, 1975.

Plamer, B. Managing Software Risk. *Medical Device & Diagnostic Industry* 19 (1997): 36–40.

Posner, K.L., Freund, P.R. Trends in Quality of Anesthesia Care Associated with Changing Staffing Patterns. Productivity and Concurrency of Case Supervision in a Teaching Hospital. *Anesthesiology* 91 (1999): 839–847.

Preboth, M. Medication Errors in Pediatric Patients. *American Family Physician* 63 (2001): 678.

Pretzer, M. Congress Backs Away from Mandatory Reporting of Medical Errors. *Medical Economics* 77 (2000): 25–26.

Purday, J.P. Monitoring During Paediatric Cardiac Anaesthesia. *Canadian Journal of Anaesthesia* 41 (1994): 818–844.

Purday, J.P., Towey, R.M. Apparent Pacemaker Failure Caused by Activation of Ventricular Threshold Test by a Magnetic Instrument Mat During General Anaesthesia. *British Journal of Anaesthesia* 69 (1992): 645–646.

Ransom, S.B., Joshi M., Nash, D. *The Healthcare Quality Book: Vision, Strategy, and Tools.* Chicago, IL: Health Administration Press, 2005.

Rao, S.N. Errors in the Treatment of Tuberculosis in Baltimore. *Journal of the American Medical Association* 283 (2000): 2502.

Rappaport, M. Human Factors Applications in Medicine. *Human Factors* 12 (1970): 25–35.

Rascona, D., Gubler, K.D., Kobus, D.A., Amundson, D., Moses, J.D. A Computerized Medical Incident Reporting System for Errors in the Intensive Care Unit: Initial Evaluation of Interrater Agreement. *Military Medicine* 166 (2001): 350–353.

Rasmussen, J. Human Error and the Problem of Causality in Analysis of Accidents. *Philosophical Transactions of the Royal Society of London—Series B: Biological Sciences* 327 (1990): 449–460.

Rastogi, A.K. High-Quality Care at Low Cost. *Medical Device & Diagnostic Industry* 19 (1997): 28–32.

Rau, G., Tripsel, S. Ergonomic Design Aspects in Interaction Between Man and Technical Systems in Medicine. *Medical Progress Through Technology* 9 (1982): 153–159.

Ray, P., Weerakkody, G. Quality of Service Management in Healthcare Organizations: A Case Study. Proceedings of the IEEE Symposium on Computer-Based Medical Systems, 1999, 80–85.

Redig, G., Swanson, M. Total Quality Management for Software Development. Proceedings of the 6th Annual IEEE Symposium on Computer-Based Medical Systems, 1993, 301–306.

Reed, L., Blegen, M.A., Goode, C.S. Adverse Patient Occurrences as a Measure of Nursing Care Quality. *Journal of Nursing Administration* 28 (1998): 62–69.

Reeter, A.K. The Role of Training in Human Factors. Proceedings of the First Symposium on Human Factors in Medical Devices, 1989, 30–31.

Reid, R.D., Christensen, M.M. Quality Healthcare – A Path Forward. Proceedings of the ASQ's 55th Annual Quality Congress, 2001, 57–63.

Rendell-Baker, L. Some Gas Machine Hazards and Their Elimination. *Anesthesia Analgesia* 88 (1977): 26–33.

Reupke, W.A., Srinivasan, R., Rigterink, P.V., Card, D.N. The Need for a Rigorous Development and Testing Methodology for Medical Software. Proceedings of the Symposium on the Engineering of Computer-Based Medical Systems, 1988, 15–20.

Revere, L., Black, K., Huq, A. Integrating Six Sigma and CQI for Improving Patient Care. *TQM Magazine* 16 (2004): 105–113.

Ribiere, V., La Salle, A.J., Khorramshahgol, R. Hospital Information Systems Quality: A Customer Satisfaction Assessment Tool. Proceedings of the 32nd Annual Hawaii International Conference on System Sciences, 1999, 140.

Richard, C., Woods, C. Operating at the Sharp Ends: The Complexity of Human Error. In *Human Error in Medicine*, edited by M.S. Bogner, 255–310. Hillsdale, NJ: Lawrence Erlbaum Associates, 1994.

Richard, J. Identifying Ways to Reduce Surgical Errors. *Journal of the American Medical Association* 275 (1996): 35.

Richards, C.F., Cannon, C.P. Reducing Medication Errors: Potential Benefits of Bolus Thrombolytic Agents. *Academic Emergency Medicine* 7 (2000): 1285–1289.

Richards, P., Kennedy, I.M., Woolf, L. Managing Medical Mishaps. *British Medical Journal* 313 (1996): 243–244.

Ridgway, M. Analyzing Planned Maintenance (PM) Inspection Data by Failure Mode and Effect Analysis Methodology. *Biomedical Instrumentation and Technology* 37 (2003): 167–169.

Riley, W.J., Densford, J.W. Processes, Techniques, and Tools: The "How" of a Successful Design Control System. *Medical Device & Diagnostic Industry* 19 (1997): 75–80.

Ritchie, J. Doctors Make Mistakes that Are Less Obvious than Lawyers' Mistakes. *British Medical Journal* 310 (1995): 1671–1672.

Robert, D.W., Robert, B.H. Preventable Deaths: Who, How Often, and Why? *Annals of Internal Medicine* 109 (1998): 582–589.

Rocha, L.S., Bassani, J.W.M. Cost Management of Medical Equipment Maintenance. Proceedings of the IEEE Annual International Conference on Engineering in Medicine and Biology, 2004, 3508–3511.

Rogers, A.S., Isreal, E., Smith, C.R. A Physician's Knowledge, Attitudes, and Behaviour Related to Reporting Adverse Drug Events. *Archives of Internal Medicine* 148 (1988): 1596–1600.

Rollins, G. Study First to Analyze Multiple Surveillance Methods for Medical Device Errors. *Report on Medical Guidelines and Outcomes Research* 15 (2004): 9–12.

Rooks, J., Zedick, J.L. The Baldrige Criteria: Managing for Quality in Healthcare. Proceedings of the 56th Annual Quality Congress, 2002, 549–550.

Rooney, J.J. Seven Steps to Improve Safety for Medical Devices, *Quality Progress* 34 (2001): 33–41.

Rose, H.B. A Small Instrument Manufacturer's Experience with Medical Instrument Reliability. Proceedings of the Annual Reliability and Maintainability Symposium, 1972, 251–254.

Rosenthal, M.M., Lloyd, S.B., eds. *Medical Mishaps: Pieces of the Puzzle*. London: Open University Press, 1999.

Rubin, S.B., Zoloth, L. Margin of Error: The Ethics of Mistakes in the Practice of Medicine. *New England Journal of Medicine* 344 (2001): 374–376.

Rudov, M.H. Professional Standards Review in Health Systems. In *Human Factors in Health Care*, edited by R.M. Pickett, T.J. Triggs, 7–47. Lexington, MA: Lexington Books, 1977.

Runciman, W.B., Helps, S.C., Sexton, E.J., Malpass, A. A Classification for Incidents and Accidents in the Health-Care System. *Journal of Quality and Clinical Practice* 18/3 (1998): 199–212.

Runciman, W.B., Webb, R.K., Lee, R. System Failure: An Analysis of 2000 Incidents Reports. *Anaesthesia and Intensive Care* 21 (1993): 684–695.

Russell, C. Human Error: Avoidable Mistakes Kill 100,000 a Year, *Washington Post*, February 18, 1992, WH7.

Sahni, A. Seven Basic Tools that Can Improve Quality, *Medical Device & Diagnostic Industry* 20 (1998): 89–96.

Santel, C., Trautmann, C., Liu, W. The Integration of a Formal Safety Analysis into the Software Engineering Process: An Example from the Pacemaker Industry. Proceedings of the Symposium on the Engineering of Computer Based-Medical Systems, 1988, 152–154.

Sayre, K., Kenner, J., Jones, P.L. Safety Models: An Analytical Tool for Risk Analysis of Medical Device Systems. Proceedings of the IEEE Symposium on Computer-Based Medical Systems, 2001, 445–454.

Scheffler, A.L., Zipperer, L. Patient Safety. Proceeding of the Symposium on Enhancing Patient Safety and Reducing Errors in Health Care, Rancho Mirage, CA, November 1998, 115–120.

Schneider, P., Hines, M.L.A. Classification of Medical Device Software. Proceedings of the Symposium on Applied Computing, 1990, 20–27.

Schreiber, P. Human Factors Issues with Anesthesia Devices. Proceedings of the First Symposium on Human Factors in Medical Devices, 1989, 32–36.

Schwartz, A.P. A Call for Real Added Value. *Medical Industry Executive* February/March (1994): 5–9.

Schwartz, R.J., Weiss, K.M., Buchanan, A.V. Error Control in Medical Data. *MD Computing* 2 (1985): 19–25.

Scott, D. Preventing Medical Mistakes. *RN Magazine* 63 (2000): 60–64.

Senders, J.W. Medical Devices, Medical Errors, and Medical Accidents. In *Human Error in Medicine*, edited by M.S. Bogner, 159–177. Hillsdale, NJ: Lawrence Erlbaum Associates, 1994.

Serig, D.I. Radiopharmaceutical Misadministrations: What's Wrong. In *Human Error in Medicine*, edited by M.S. Bogner, 179–195. Hillsdale, NJ: Lawrence Erlbaum Associates, 1994.

Sexton, J.B., Thomas, E.J., Helmreich, R.L. Error, Stress, and Teamwork in Medicine and Aviation: Cross Sectional Surveys. *British Medical Journal* 320 (2000): 745–749.

Shanghnessy, A.F., Nickel, R.O. Prescription-Writing Patterns and Errors in a Family Medicine Residency Program. *Journal of Family Practice* 29 (1989): 290–296.

Shaw, R. Safety-Critical Software and Current Standards Initiatives. *Computer Methods and Programs in Biomedicine* 44 (1994): 5–22.

Shea, C.E., Battles, J.B. A System of Analyzing Medical Errors to Improve GME Curricula and Programs, *Academic Medicine* 76 (2001): 125–133.

Shelton, N. To Err Is Human. *Skeptical Inquirer* 20 (1996): 21–22.

Shepherd, M. A Systems Approach to Medical Device Safety. Monograph, Association for the Advancement of Medical Instrumentation, Arlington, Virginia, 1983.

Shepherd, M., Brown, R. Utilizing a Systems Approach to Categorize Device-Related Failures and Define User and Operator Errors. *Biomedical Instrumentation and Technology* November/December (1992): 461–475.

Shepherd, M., Painter, F.R., Dyro, J.F., Baretich, M.F. Identification of Human Errors During Device-Related Accident Investigations. *IEEE Engineering in Medicine and Biology* 23 (2004): 66–72.

Shephard, M.D. *Shepherd's System for Medical Device Incident Investigation*. Brea, CA: Quest Publishing Company, 1992.

Sheridan, T. B., Thompson J.M. People Versus Computers in Medicine. In *Human Error in Medicine*, edited by M.S. Bogner, 141–159. Hillsdale, NJ: Lawrence Erlbaum Associates, 1994.

Shinn, J.A. Root Cause Analysis: A Method of Addressing Errors and Patient Risk. *Progress in Cardiovascular Nursing* 15 (2000): 25–25.

Short, T.G., O'Regan, A., Oh, T.E. Critical Incident Reporting in an Anaesthetic Department Assurance Programme. *Anaesthesia* 47 (1992): 3–7.

Simonaitis, D.F., Anderson, R.T., Kaye, M.P. Reliability Evaluation of a Heart Assist System. Proceedings of the Annual Reliability and Maintainability Symposium, 1972, 233–241.

Smirnov, I.P., Shneps, M.A. Medical System Engineering. *Proceedings of the IEEE 57* 11 (1969): 1869–1879.

Smith, C.E., Peel, D. Safety Aspects of the Use of Microprocessors in Medical Equipment. *Measurement and Control* 21 (1988): 275–276.

Smith, J. Study into Medical Errors Planned for the UK. *British Medical Journal* 319 (1999): 1091–1092.

Smith, L.S. Documenting Known or Suspected Medical Failure. *Nursing* 34 (2004): 28–29.

Sommer, T.J. Telemedicine: A Useful and Necessary Tool to Improving Quality of Healthcare in the European Union. *Computer Methods and Programs in Biomedicine* 48 (1995): 73–77.

Souhrada, L. Human Error Limits MRI's Quality Potential. *Hospitals* 63 (1989): 38.

Souhrada, L. Man Meets Machine: Buying Right Can Reduce Human Error. *Materials Management in Health Care* 4 (1995): 20–22.

Spath, P.L. Medical Errors: Root Cause Analysis. *OR Manager* 14 (1998): 40–41.

Spencer, F.C. Human Error in Hospitals and Industrial Accidents: Current Concepts. *Journal of the American College of Surgeons* 191 (2000): 410–418.

Spice, C. Misdiagnosis of Ventricular Tachycardia. *Lancet* 354 (1999): 2165–2169.

Spicer, J. How to Measure Patient Satisfaction. *Quality Progress* 35 (2002): 97–98.

Staender, S., Davies, J., Helmreich, B., Sexton, B., Kaufmann, M. The Anaesthesia Critical Incident Reporting System: An Experience Based Database. *International Journal of Medical Informatics* 47 (1997): 87–90.

Stahlhut, R.W. To Err Is Human: Human Error in Medicine Is Common, Inevitable and Manageable. *Medical Device & Diagnostic Industry* 19 (1997): 13–14.

Stahlhut, R.W., Gosbee, J.W., Gardner-Bonneau, D.J. A Human-Centered Approach to Medical Informatics for Medical Students, Residents, and Practicing Clinicians. *Academic Medicine* 72 (1997): 881–887.

Stanley, P.E. Monitors that Save Lives Can Also Kill. *Modern Hospital* 108 (1967): 119–121.

Stanton, N.A., Stevenage, S.V. Learning to Predict Human Error: Issues of Acceptability, Reliability and Validity. *Ergonomics* 41 (1998): 1737–1756.

Stein, P.E. Reliability and Performance Criteria for Electromedical Apparatus. Proceedings of the Annual Reliability and Maintainability Symposium, 1973, 89.

Steindel, B. Quality Control in the Practice of Medicine. Proceedings of the 11th Annual West Coast Reliability Symposium, 1970, 197–202.

Steward, S. Making Device Software Truly Trustworthy. *Medical Device & Diagnostic Industry* 20 (1998): 86–89.

Stoecklein, M. ASQ's Role in Healthcare. *Quality Progress* 36 (2003): 90–91.

Swartz, E.M. Product Liability, Manufacturer Responsibility for Defective or Negligently Designed Medical and Surgical Instruments. *De Paul Law Review* 18 (1969): 348–407.

Tarcinale, M.A. Patient Classification: Cutting the Margin of Human Error. *Nursing Management* 17 (1986): 49–51.

Taxis, K., Barber, N. Ethnographic Study of Incidence and Severity of Intravenous Drug Errors. *British Medical Journal* 326 (2003): 684–687.

Taylor, E.F. Reliability, Risk and Reason in Medical Equipment. Proceedings of the 5th Annual Meeting of the Association for the Advancement of Medical Instrumentation, March 23, 1970, 1–5.

Taylor, E.F. The Effect of Medical Test Instrument Reliability on Patient Risks. Proceedings of the Annual Symposium on Reliability, 1969, 328–330.

Taylor, E.F. The Impact of FDA Regulations on Medical Devices. Proceedings of the Annual Reliability and Maintainability Symposium, 1980, 8–10.

Taylor, E.F. The Reliability Engineer in the Health Care System. Proceedings of the Reliability and Maintainability Symposium, 1972, 245–248.

Taylor, G. What Is Minimal Monitoring, Essential Non-invasive Monitoring. In *Anaesthesia*, edited by Gravenstein, J.S., Newbower, R.S., Ream, A.K., Smith, N.T., Barden, J., 263–267. New York: Grune & Stratton Publishers, 1980.

Tecca, M.B., Weitzner, W.M. Measuring Performance in Home Health from Internal and External Vantage Points. Proceedings of the 9th Annual Quest for Quality and Productivity in Health Services, 1997, 27–38.

Thibeault, A. Documenting a Failure Investigation. *Medical Device & Diagnostic Industry* 19 (1997): 14–15.

Thibeault, A. Handling Reports of Product Use in Incidents Causing Injury or Death. *Medical Device & Diagnostic Industry* 20 (1998): 91–96.

Thierry, G., Dhainaut, J.F., Joseph, T., Journois, D. Iatrogenic Complications in Adult Intensive Care Units: A Prospective Two-Center Study. *Critical Care Medicine* 21 (1993): 40–52.

Thomas, E.J., Sherwood, G.D., Adams-McNeill, J. Identifying and Addressing Medical Errors in Pain Mismanagement. *Joint Commission Journal on Quality Improvement* 27 (2001): 191–199.

Thomas, E.J., Studdert, D.M., Runciman, W.B., Webb, R.K., Sexton, E.J., Wilson, R.M., et al. A Comparison of Iatrogenic Injury Studies in Australia and the USA. I: Context, Methods, Casemix, Population, Patient and Hospital Characteristics. *International Journal for Quality in Health Care* 12 (2000): 371–378.

Thompson, C.N.W. Model of Human Performance Reliability in Health Care System. Proceedings of the Annual Reliability and Maintainability Symposium, 1974, 335–339.

Thompson, P.W. Safer Design of Anaesthesia Equipment. *British Journal of Anaesthesiology* 59 (1987): 913–921.

Thompson, R.C. Fault Therapy Machines Cause Radiation Overdoses. *FDA Consumer* 21 (1987): 37–38.

Truby, C. Quality and Productivity: Then and Now. *Medical Device & Diagnostic Industry* 21 (1999): 104–108.

Turk, A.R., Poulakos, E.M. Practical Approaches for Healthcare: Indoor Air Quality Management, Energy Engineering. *Journal of the Association of Energy Engineering* 93 (1996): 12–79.

Turley, J.P., Johnson, T.R., Smith, D.P., Zhang, J., Brixey, J.J. Operating Manual-Based Usability Evaluation of Medical Devices: An Effective Patient Safety Screening Method. *Joint Commission Journal on Quality and Patient Safety* 32 (2006): 214–220.

Van Cott, H. Human Errors: Their Causes and Reduction. In *Human Error in Medicine*, edited by M.S. Bogner, 53–91. Hillsdale, NJ: Lawrence Erlbaum Associates, 1994.

Van Grunsven, P.R. Criminal Prosecution of Health Care Providers for Clinical Mistakes and Fatal Errors: Is Bad Medicine a Crime? *Journal of Health and Hospital Law* 29 (1996): 107.

Varricchio, F. Another Type of Medication Error. *Southern Medical Journal* 93 (2000): 834.

Vincent, C., Neale, G., Woloshynowych, M. Adverse Events in British Hospitals: Preliminary Retrospective Record Review. *British Medical Journal* 322 (2001): 517–519.

Vincent, C., Taylor, S., Chapman J.E. How to Investigate and Analyze Clinical Incidents: Clinical Risk Unit and Association of Litigation and Risk Management Protocol. *British Medical Journal* 320 (2000): 777–778.

Vroman, G., Cohen, I., Volkman, N. Misinterpreting Cognitive Decline in the Elderly: Blaming the Patient. In *Human Error in Medicine*, edited by M.S. Bogner, 93–122. Hillsdale, NJ: Lawrence Erlbaum Associates, 1994.

Wadsworth, H.M. Standards for Tools and Techniques. *Transactions of the ASQC Annual Conference* (1994): 882–887.

Walfish, S. Using Statistical Analysis in Device Testing. *Medical Device & Diagnostic Industry* 19 (1997): 36–42.

Wallace, D. R., Kuhn, R. D. Lessons from 342 Medical Device Failures. In *Information Technology Laboratory*. Gaithersburg, MD: National Institute of Standards and Technology, 2000, 25–31.

Walters, D., Jones, P. Value and Value Chains in Healthcare: A Quality Management Perspective. *TQM Magazine* 13 (2001): 319–333.

Wang, B., Eliason, R.W., Vanderzee, S.C. Global Failure Rate: A Promising Medical Equipment Management Outcome Benchmark. *Journal of Clinical Engineering* 31 (2006): 145–151.

Ward, J.R., Clarkson, P.J. An Analysis of Medical Device-Related Errors: Prevalence and Possible Solutions. *Journal of Medical Engineering and Technology* 28 (2004): 2–21.

Waynant, R.W. Quantitative Risk Analysis of Medical Devices: An Endoscopic Imaging Example. Proceedings of the International Society for Optical Engineering Conference, 1995, 237–245.

Wear, J.O. Maintenance of Medical Equipment in the Veterans Administration. Proceedings of the Third Annual Meeting of the Association for the Advancement of Medical Instrumentation, July 1968, 10–14.

Wear, J.O. Technology and the Future of Medical Equipment Maintenance. *Health Estate* 53 (1999): 12–19.

Wears, R.L., Janiak, B., Moorhead, J.C., Kellermann, A.L., Yeh, C.S., Rice, M.M., et al. Human Error in Medicine: Promise and Pitfalls, Part 1. *Annals of Emergency Medicine* 36 (2000): 58–60.

Wears, R.L., Janiak, B., Moorhead, J.C., Kellermann, A.L., Yeh, C.S., Rice, M.M., et al. Human Error in Medicine: Promise and Pitfalls, Part 2. *Annals of Emergency Medicine* 36 (2000): 142–144.

Wears, R.L., Leape L.L. Human Error in Emergency Medicine. *Annals of Emergency Medicine* 34 (1999): 370–372.

Webb, R.K., Russell, W.J., Klepper, I. Equipment Failure: An Analysis of 2000 Incidents Report. *Anaesthesia and Intensive Care* 21 (1993): 673–677.

Webster, S.A. Technology Helps Reduce Mistakes on Medications. *Detroit News*, February 6, 2000, A2.

Weese, D.L., Buffaloe, V.A. Conducting Process Validations with Confidence. *Medical Device & Diagnostic Industry* 20 (1998): 107–112.

Weide, P. Improving Medical Device Safety with Automated Software Testing. *Medical Device & Diagnostic Industry* 16 (1994): 66–79.

Weilgart, S.N., Ship, A.N., Aronson, M.D. Confidential Clinican-Reported Surveillance of Adverse Events Among Medical Inpatients. *Journal of General Internal Medicine* 15 (2000): 470–477.

Weinberg, D.I., Artley, J.A., Whalen, R.E., McIntosh, M.T. Electrical Shock Hazards in Cardiac Catheterization. *Circulation Research* 11 (1962): 1004–1011.

Weingart, S.N., Wilkson, R.S., Gibberd, R.W. Epidemiology of Medical Error. *British Medical Journal* 320 (2000): 774–776.

Weinger, M.B. Anesthesia Equipment and Human Error. *Journal of Clinical Monitoring & Computing* 15 (1999): 319–323.

Weinger, M.B., Englund, C.E. Ergonomic and Human Factors Affecting Anesthetic Vigilance and Monitoring Performance in the Operating Room Environment. *Anesthesiology* 73 (1990): 995–1021.

Welch, D.L. Human Error and Human Factors Engineering in Health Care. *Biomedical Instrumentation & Technology* 31 (1997): 627–631.

Welch, D.L. Human Factors Analysis and Design Support in Medical Device Development. *Biomedical Instrumentation & Technology* 32 (1998): 77–82.

Welch, D.L. Human Factors in the Health Care Facility. *Biomedical Instrumentation &Technology* 32 (1998): 311–316.

Whalen, R.E., Starmer, C.F., McIntosh, H.D. Electrical Hazards Associated with Cardiac Pacemaking. *Transactions of the New York Academy of Sciences* 111 (1964): 922–931.

Widman, L.E., Tong, D.A. EINTHOVEN and Tolerance for Human Error: Design Issues in Decision Support System for Cardiac Arrhythmia Interpretation. Proceedings of the AMIA Annual Fall Symposium, 1986, 224–228.

Wiklund, M.E. Human Error Signals Opportunity for Design Improvement. *Medical Device & Diagnostic Industry* 14 (1992): 57–61.

Wiklund, M.E. *Medical Device and Equipment Design.* Buffalo Grove, IL: Interpharm Press, 1995.

Wiklund, M.E. Making Medical Device Interfaces More User Friendly. *Medical Device & Diagnostic Industry* 20 (1998): 177–183.

Wiklund, M.E., Weinger, M.B., Pantiskas, C., Carstensen, P. Incorporating Human Factors into the Design of Medical Devices. *Journal of the American Medical Association* 280 (1998): 1484.

Wilkins, R.D., Holley, L.K. Risk Management in Medical Equipment Management. Proceedings of the 20th Annual International Conference of the IEEE Engineering in Medicine and Biology Society, 1998, 3343–3345.

Williamson, J.A., Webb, R.K., Sellen, A. Human Failure: An Analysis of 2000 Incidents Reports. *Anaesthsia and Intensive Care* 21 (1993): 678–683.

Willis, G. Failure modes and effects analysis in clinical engineering. *Journal of Clinical Engineering* 17 (1992): 59–63.

Wilson, C.J. ME: A Survey of Anesthetic Misadventures. *Anaesthesia* 36 (1981): 933–936.

Winger, J., Bray, T., Halter, P. Lower Health Costs from High Reliability. Proceedings of the Annual Reliability and Maintainability Symposium, 1979, 203–210.

Winterberg, L. Quality Improvement in Healthcare. Proceedings of the ASQ's 55th Annual Quality Congress, 2001, 352–353.

Wood, B.J. Software Risk Management for Medical Devices. *Medical Device & Diagnostic Industry* 21 (1999): 139–145.

Wood, B.J., Ermes, J.W. Applying Hazard Analysis to Medical Devices, Part II. *Medical Device & Diagnostic Industry* 15 (1993): 58-64.

Wright, D. Critical Incident Reporting in an Intensive Care Unit: 10 Years' Experience, Intensive Therapy Unit. Report, Western General Hospital, Edinburgh, Scotland, 1999.

Wu, A.W. Handling Hospital Errors: Is Disclosure the Best Defence? *Annals of Internal Medicine* 131 (1999): 970-972.

Wu, A.W. Medical Error: The Second Victim. *British Medical Journal* 320 (2000): 726-727.

Wu, A.W., Folkman, S., Mc Phee, S.J. How House Officers Cope with Their Mistakes. *Western Journal of Medicine* 159 (1993): 565-569.

Yi, G., Johnson, T.R., Patel, Y.L. Research on Basic Element for Medical Equipment Maintenance Satisfaction Under Requirement Theory. *Journal of Services Operations and Informatics* 1 (2002): 165-173.

Zane, M. Patient Care Appraisal. Proceedings of the Annual Reliability and Maintainability Symposium, 1976, 84-91.

Zhang, J., Xing-San, Q., Wei, D. Using Usability Heuristics to Evaluate Patient Safety of Medical Devices. *Journal of Biomedical Informatics* 36 (2003): 23-30.

Zimmerman, J.E., Knaus, W.A., Wagner, D.P., Draper, E.A., Lawrence, D.E. The Range of Intensive Care Services Today. *Journal of the American Medical Association* 246 (1981): 2711-2716.

Zimmerman, J.E., Shortell, S.M., Rousseau, D.M., Duffy, J., Gillies, R.R. Improving Intensive Care: Observations Based on Organizational Case Studies in Nine Intensive Care Units: A Prospective Multicenter Study. *Critical Care Medicine* 21 (1983): 1443-1451.

Index

A

B

C

D

E

F

G

H

I

J

L

M

N

O

P

Q

R

S

T

U

V

W

X

Z

www.ingramcontent.com/pod-product-compliance
Ingram Content Group UK Ltd.
Pitfield, Milton Keynes, MK11 3LW, UK
UKHW021612240425
457818UK00018B/510

* 9 7 8 0 3 6 7 3 8 7 7 5 4 *